SpringerBriefs in Environmental Science

W0116095

SpringerBriefs in Environmental Science present concise summaries of cutting-edge research and practical applications across a wide spectrum of environmental fields, with fast turnaround time to publication. Featuring compact volumes of 50 to 125 pages, the series covers a range of content from professional to academic. Monographs of new material are considered for the SpringerBriefs in Environmental Science series.

Typical topics might include: a timely report of state-of-the-art analytical techniques, a bridge between new research results, as published in journal articles and a contextual literature review, a snapshot of a hot or emerging topic, an in-depth case study or technical example, a presentation of core concepts that students must understand in order to make independent contributions, best practices or protocols to be followed, a series of short case studies/debates highlighting a specific angle.

SpringerBriefs in Environmental Science allow authors to present their ideas and readers to absorb them with minimal time investment. Both solicited and unsolicited manuscripts are considered for publication.

More information about this series at http://www.springer.com/series/8868

Marzia Traverso • Luigia Petti
Alessandra Zamagni

Editors

Perspectives on Social LCA

Contributions from the 6th International
Conference

 Springer

Editors
Marzia Traverso
Institute for Sustainability in Civil
Engineering
RWTH Aachen University
Aachen, Germany

Luigia Petti
Dipartimento di Economia
University of Chieti-Pescara
Pescara, Italy

Alessandra Zamagni
Ecoinnovazione srl
Spin-off ENEA
Bologna, Italy

ISSN 2191-5547   ISSN 2191-5555   (electronic)
SpringerBriefs in Environmental Science
ISBN 978-3-030-06564-5   ISBN 978-3-030-01508-4   (eBook)
https://doi.org/10.1007/978-3-030-01508-4

Library of Congress Control Number: 2019933900

This Springer imprint is published by the registered company Springer Nature Switzerland AG.
The registered company address is: Gewerbestrasse 11, 6330 Cham, Switzerland

Preface

The Social Life Cycle Assessment (S-LCA) is officially recognised to be part of Life Cycle Thinking (LCT), and since May 2018, it is again a topic under the umbrella of the UN Environment Life Cycle Initiative activities. In fact, the current guidelines, published by UNEP Life Cycle Initiative in 2009, are under revision, in the framework of a project sponsored by the Life Cycle Initiative, and their launch with relative pilot projects, is expected at LCM2019 Conference in September 2019. In the last 10 years, several S-LCA developments and implementations have been carried out, increasing the importance of the S-LCA in both private and public sectors. Given the economic crisis, attention has been brought on the social component of the sustainability both in Europe and in the developed countries more in general, highlighting that the management of the social issues is not only a need but also an opportunity, because it further qualifies the product/service on the market. In addition, it is an opportunity to reward those organisations that are already creating social value through the reinvestment of their profits into cultural and social initiatives for the community. In other words, organisations can be the leverage for social value creation, and their competitiveness can benefit from it. For this reason, the interest of the policy-makers has increased in order to identify the positive and negative social hotspots generated by a product or a company in different local contests.

The S-LCA conferences have today reached the sixth version and it is today an international event that allows experts and non-experts from the academy, industry and policy to meet and exchange on this topic and to discuss its challenges. Several improvements and more interest from stakeholders outside the scientific community have been registered since the first seminar held in Lyngby at the Technical University of Denmark on 31 May 2010, promoted by Dr. Louise Camilla Dreyer.

The aim of the sixth International Conference on S-LCA People&Places4Partnership is to discuss about the key role of S-LCA as a decision-making tool in the definition of strategies for social sustainability, thus supporting both public and private businesses in making more informed decisions. In this conference, three sessions have been organised: scientific presentations, industry sessions and a policy workshop to underline the necessity to discuss the potentials, challenges and gaps of S-LCA at different levels. The conference has registered more than 130 participants and more than 60 contributions, whose abstracts are reported in the conference proceedings. A limited number of full papers have been selected to be published in this book to represent the state of the art and some of the current initiatives and implementations of S-LCA. The book starts with few examples on further developments of the S-LCA phases, in particular: the definition of the functional unit, in the framework of the goal and scope phase (Arzoumanidis et al. 2018), and the definition and development of impact pathway and weighting approaches in the impact assessment phase (Weidema 2018, Di Cesare et al. 2018, Benoit-Norris et al. 2018 and Breno et al. 2018). Then, some examples of alternative approaches are presented, developed in the industrial context to measure the social impact (Baumann et al. 2018, Saling et al. 2018, and Vuaillat et al. 2018). Finally, three contributions are focusing on practical implementations of S-LCA to different activity sectors: waste management (Ibañez-Forés et al. 2018), automotive components (Zanchi et al. 2018) and agriculture system (Frank 2018).

Aachen, Germany   M. Traverso
Bologna, Italy   A. Zamagni
Pescara, Italy   L. Petti

Contents

Chapter 1
Functional Unit Definition Criteria in Life Cycle Assessment and Social Life Cycle Assessment: A Discussion

Ioannis Arzoumanidis, Manuela D'Eusanio, Andrea Raggi, and Luigia Petti

Abstract The definition of a Functional Unit (FU) is essential for building and modelling a product system in Life Cycle Assessment (LCA). A FU is a quantified description of the function of a product that serves as the reference basis for all calculations regarding impact assessment. A function may be based on different features of the product under study, such as performance, aesthetics, technical quality, additional services, costs, etc. Whilst the FU definition is typical in LCA, this does not seem to be a common practice in Social Life Cycle Assessment (S-LCA), even though a FU definition is required. Unlike LCA, where quantitative data are mainly collected and processed, the assessment of the social and socio-economic impacts in S-LCA is based on a prevalence of qualitative and semi-quantitative data, a fact that renders the assessment to be somehow unfriendly. Moreover, whilst in LCA a product-oriented approach is typical, S-LCA tends to be a business-oriented methodology, where the emphasis of the social assessment lies on the behaviour of the organisations that are involved in the processes under study rather than on the function that is generated by a product. Indeed, several S-LCA case studies were found in the literature in which the FU is not discussed, let alone defined. The objective of this article is to contribute to analysing the criteria used for the definition of a FU in LCA and verifying whether these criteria can be suitable for S-LCA case studies applications. For this reason, a literature review was carried out on LCA in order to identify whether and how this issue has been tackled with so far. In addition, a second literature review was performed in order to verify how the FU has been introduced in the framework of the S-LCA methodology. Finally, an investigation of the analysis results, in terms of the selected FU, is proposed in view of an ever-growing need for a combination of the LCA and S-LCA methodologies into a broader Life Cycle Sustainability Assessment (LCSA).

I. Arzoumanidis (✉) · M. D'Eusanio · A. Raggi · L. Petti
Department of Economic Studies (DEc), University "G. d'Annunzio", Pescara, Italy
e-mail: i.arzoumanidis@unich.it

1

M. Traverso et al., *Perspectives on Social LCA*, SpringerBriefs in Environmental Science, https://doi.org/10.1007/978-3-030-01508-4_1

1.1 Introduction

With the definition of Sustainable Development at the Conference of Rio in 1992, sustainability has become an inseparable part of the core decision-making processes and a strategic objective for business and governance. A product may be considered to be sustainable if there is an equilibrium between the three dimensions: economic, environmental and social [1, 2]. In order for the sustainability of a product, an organisation or a process to be assessed, Life Cycle Thinking (LCT) methods and tools can be implemented. Amongst these, LCA focuses on the environmental issues, whilst S-LCA analyses the social ones. Both methodologies draw from the ISO 14040:2006 framework [3], but have different application characteristics [4]. Indeed, whilst LCA is based on the physical flows of a product system [3], S-LCA considers the behaviour of the companies involved in the related processes [5]. Moreover, the nature of the assessed impacts and the presence of both qualitative and semi-qualitative data in S-LCA, render the assessment to be strongly context-related [6]. On the other hand, LCA uses quantitative product-related data [3]. As already known, the framework of the two methodologies consists in the following phases: (1) Goal and Scope Definition (GSD); (2) Life Cycle Inventory (LCI); (3) Life Cycle Impact Assessment (LCIA); (4) Interpretation [3, 7].

This article focuses on the first phase of the LCT methodologies (GSD) and specifically on the FU definition. ISO 14040:2006 defines FU as the "quantified performance of a product system for use as a reference unit" [3; p. 4]. This definition is also adopted by the S-LCA methodology [7], the guidelines of which explicitly refer to the ISO 14040:2006 standard. The FU describes and quantifies the features of a product (functionality, aspect, stability, durability, ease of maintenance, etc.), which are market-driven [8]. The objective of this study is to analyse the FU definition and identification in LCA and S-LCA, in order to highlight differences and similarities and to ascertain whether it is reasonable and possible to extend the typical LCA FU definition criteria to the social evaluation of a product. For this purpose, the scientific literature for both methodologies to identify the criteria for defining FU was analysed. Since the two methodologies present a different development level, the two literature reviews were performed using partially dissimilar approaches, as described in Sects. 1.3–1.4. This article is structured as follows: Sects. 1.2–1.4 describe the literature review methodologies and Sect. 1.5 the results for LCA and S-LCA. In Sect. 1.6, the elements of similarity and differentiation between LCA and S-LCA regarding the FU identification are discussed, and in Sect. 1.7, some conclusions are drawn.

1.2 Methodology

Given that the two analysed methodologies present a different level of development, the two literature reviews were performed using partially dissimilar approaches. The relevant search strategies will be described in Sects. 1.3 and 1.4; however, both searches were carried out using the same research engine (EBSCO Discovery Service available at the Univ. "G. d'Annunzio" Library's website) [9] and without imposing any initial time limit (the end of the time interval was set at the end of October 2017). The review was performed by searching for words such as "functional unit", "function*", "reference flow", "reference" and "flow" within the identified articles. Finally, in order to render the two analyses more homogeneous, the same sectorial categorisation was used, i.e., based on the NACE (*Nomenclature statistique des Activités économiques dans la Communauté Européenne*) codes [10].

1.3 Literature Review on LCA

LCA evaluates the environmental impacts throughout the entire life cycle of a product and is an ever more applied methodology for improving the environmental performance of products and services [11]. Given the great number of published case studies, concerning different sectors, the literature review was limited to considering only case studies and methodological reviews. The search used the terms ("LCA" OR "Life Cycle Assessment") AND "review" in the field of the title (of the articles). In this way, all possible sectors of products and services were considered. 326 results initially emerged; these were subsequently restricted by means of a screening procedure to 111 review articles for various sectors (excluding those papers related to phases other than GSD). Fig. 1.1 presents the distribution of the publications per sector, whilst Fig. 1.2 the temporal one.

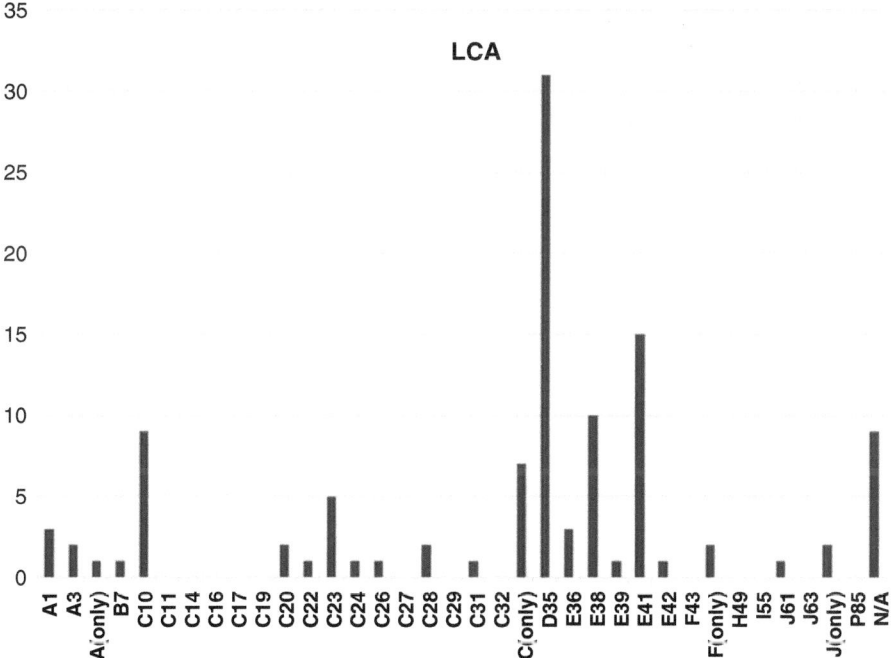

Fig. 1.1 LCA – Distribution of the reviewed publications per sector (NACE code)

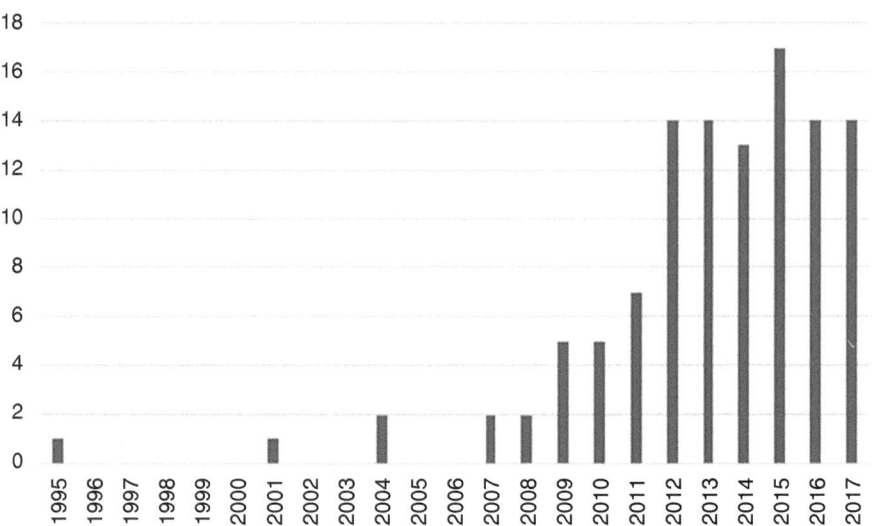

Fig. 1.2 LCA – Temporal distribution of the reviewed publications

1.4 Literature Review on S-LCA

The S-LCA literature research was conducted using the terms "Social Life Cycle Assessment", "Social LCA", "S-LCA" and "S-LCA" and the OR operator, resulting in 7129 articles. Given the high number of results, a filter was applied in the "subject" field, thus considering only the articles that dealt with "Social Life Cycle Assessment", "S-LCA" and "social impacts", thus arriving at 133 results. Subsequently, the articles were divided into three macro-areas: methodological, reviews and case studies. The literature review showed that the publications distribution by type is made up of 52.63% of case-studies, followed by methodological articles (34.59%) and reviews (12.78%). Here, only S-LCA case studies were considered in order to identify the FU selection criteria. The first S-LCA studies emerged in 2006 (Fig. 1.4). The frequency of the case studies per year shows that since 2009 there has been an increase in S-LCA articles, probably following the publication of the Guidelines [7]. Figure 1.3 presents the distribution of the publications per sector, whilst Fig. 1.4 the temporal one.

1.5 Results

The results of the findings are presented hereafter, whilst a detailed description of the FU quantities used in the various sectors, along with their frequency of presence, is presented in Table 1.1. Please note that in order to categorise the identified sectors, these were brought to the first level of detail of the NACE codes (e.g., C10.1.5 was

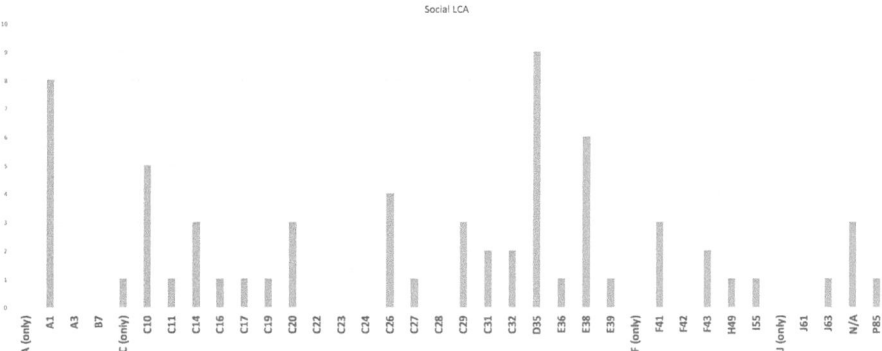

Fig. 1.3 S-LCA – Distribution of the reviewed publications per sector (NACE codes)

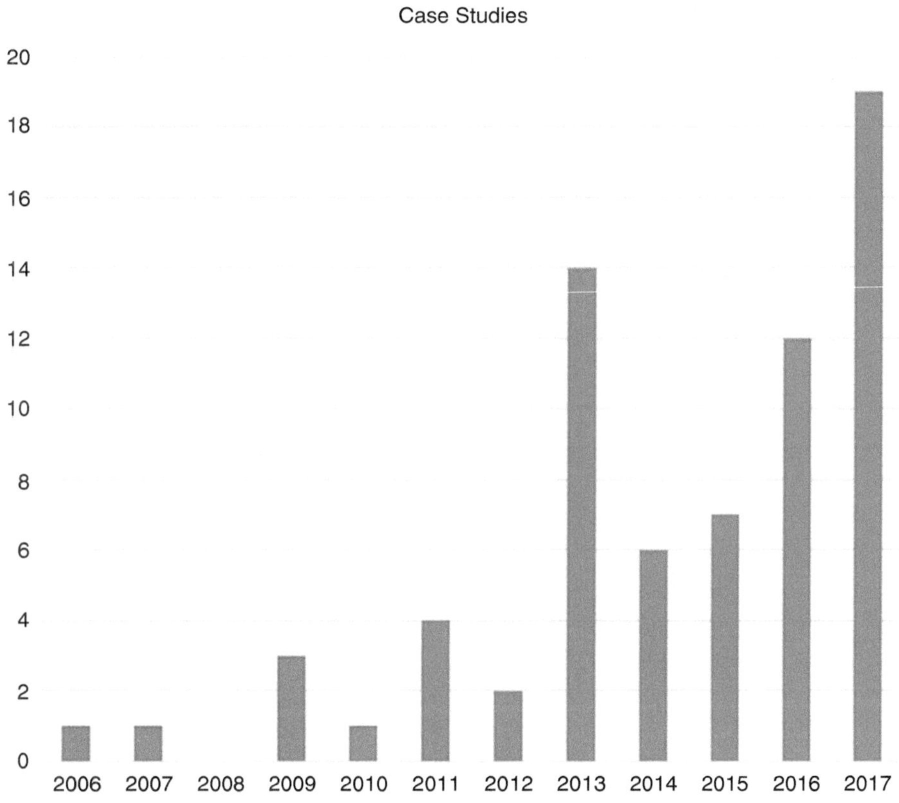

Case Studies

Fig. 1.4 S-LCA- Temporal distribution of the reviewed publications

Table 1.1 Summary of the review results (the number of publications found for that quantity, if more than one, is in parenthesis)

Sectors	NACE code	No. of identified articles		FU quantity	
		LCA	S-LCA	LCA	S-LCA
Agriculture, forestry and fishing	A1	3	8	Mass (3); area (3); energy (2); product unit; economic value; volume	N/A (3); mass (3); product unit, area
	A3	2		Mass (2); calorific value	
	A (only)	1		N/A	
Mining and quarrying	B7	1		N/A	
Manufacturing	C10	9	5	Mass (7); volume (6); nutritional value (3); area (3); economic value (2); energy; profit; N/A (2)	Mass (4); N/A
	C11		1		Volume
	C14		3		N/A (2); economic value
	C16		1		Product unit
	C17		1		Person time
	C19		1		N/A
	C20	2	3	Mass; N/A	Mass (3)
	C22	1		N/A	
	C23	5		Volume (3); mass (2); distance (2); product unit (2); N/A	
	C24	1		N/A	
	C26	1	4	N/A	Product unit (2); N/A (2)
	C27		1		N/A
	C28	2		N/A; product unit; area; distance, volume	
	C29		3		Product unit (2); N/A
	C31	1	2	Product unit	N/A; product unit
	C32		2		Product unit (2)
	C (only)	7	1	N/A (3); mass (3); product unit (2); volume; energy; environmental impact	Product unit
Electricity, gas, steam and air conditioning supply	D35	31	9	Energy (20); mass (17); area (10); distance (8); N/A (8); volume (6); product unit (3); environmental impact (2); time (2); service; yield; calorific value	N/A (3); mass (3); volume; distance; area

(continued)

Table 1.1 (continued)

Sectors	NACE code	No. of identified articles LCA	S-LCA	FU quantity LCA	S-LCA
Water supply; sewerage; waste management and remediation activities	E36	3	1	Volume (3)	Mass
	E38	10	6	Mass (4); N/A (3); volume (3); environmental impact (2); quantity; quality	Mass (4); N/A; person time
	E39	1	1	Mass; volume	Mass
Construction	F41	15	3	Area (12); product unit (7); mass (4); energy (4); volume (3); time (2); value (2); insulating value (2); N/A (2); environmental impact	Product unit; N/A; mass
	F42	1		N/A	
	F43		2		Area (2)
	F (only)	2		Mass; N/A	
Transporting and storage	H49		1		Mass
Accommodation and food service activities	I55		1		Time
Information and communication	J61	1		N/A	
	J63		1		Person time
	J (only)	2		Product unit; N/A	
Education	P85		1		N/A
Not available/not identifiable	N/A	9	3	N/A (6); mass (3); energy; volume	N/A (3)

brought to C10), whilst the zero-level codes -- e.g., A (only) -- refer to sectors for which the first level of detail was not available within the reviewed articles.

1.5.1 LCA

As expected, the literature review showed an increase in published reviews in recent years [12]. The most cited sector in the analysed reviews is the energy-related one (sector D35), followed by the construction of buildings (sector F41). Some of the analysed reviews provided details for the different identified FUs (Table 1.1). In 76 out of 111 review articles (68.47%) the FU is discussed and defined in different ways (e.g., for the manufacturing of food products (sector C10), the FU is identified in terms of mass, product unit, energy, area, volume, nutritional or economic value, etc.), whilst for the remaining 31.53% no FU definition was given (e.g., for sector C10, the FU was not examined at all in two reviews). Moreover, a detailed description of the FU was provided only in 59.46% of the articles examined (even if, not always in an adequate way), whilst an attempt to give a description of the function of the product was provided in even fewer cases (10.81% of the articles). Furthermore, Table 1.1 shows the prevailing FU quantities for each sector. Regarding the energy-related sector (D35), the most commonly used quantity to define the FU is obviously energy, followed by mass (e.g., of a specific fuel). In general, the most used quantity is mass, followed by energy, volume and area (Table 1.1). Finally, whilst for some sectors, specific FUs are found (e.g., the economic value for the manufacturing of food products and insulating value for the buildings sector), it is noted that most of the defined FUs (e.g., mass, volume and energy) are common for several sectors.

1.5.2 S-LCA

FU identification is considered to construct and model the product-system and thus identify the context and the stakeholders involved in the study itself [7]. Since S-LCA evaluates the social aspects of the products, it uses mainly qualitative data and indicators, which, in the LCIA phase, do not allow an immediate link of the results to the FU (ibid.). The most cited sector in the S-LCA analysed articles is the energy-related one (sector D35), followed by agriculture (A1). The analysed papers provided details for the different identified FUs (Table 1.1). The review showed a non-negligible presence of case studies where a FU was not identified, let alone discussed (24.72%), whilst for the remaining (75.28%) the FU was taken into consideration. In the papers where the FU is discussed, the most common FU refers to mass (23 papers), followed by the product unit (9 papers) (Table 1.1). The choice of both mass and product unit as a quantity is found in different sectors (e.g., agriculture, manufacturing). On the other hand, mass was selected for the energy, waste management and transport sectors, whilst product unit was selected for the construction sector. Moreover, Table 1.1 shows that the most commonly analysed sector is manufacturing, which includes different specific sub-sectors i.e., food, electronics, textile products, etc.

1.6 Discussion

This analysis showed that the recurring economic sectors in the case studies are different for LCA and S-LCA. A comparison of the different FUs was possible only between the sectors present in both cases (11 sectors). For instance, for the manufacture of food products, a mass-based FU is prevalent in both methodologies (Fig. 1.1). Table 1.1 shows the quantities mainly used. It can be noted that, for 6 out of 11 sectors, the same quantity is used: mass in 5 sectors (A1 – crop and animal production, hunting and related service activities; C10 – manufacture of food products; C20 – manufacture of chemicals and chemical products; E38 – waste collection, treatment and disposal activities; materials recovery; E39 – remediation activities and other waste management services); product unit for one sector (sector C31 – manufacture of furniture). On the other hand, for four sectors (C only – manufacturing; D35 – electricity, gas, steam and air conditioning supply; E36 – water collection, treatment and supply; F41 – construction of buildings) different FUs are used. Furthermore, this comparison cannot be made for the C26 sector (manufacture of computer, electronic and optical products) as no reference to the FU was found in the LCA review.

Regarding the presence of a FU definition, although it is an important aspect for the individual case studies of LCA, it does not seem to have received the same attention in the review articles. Indeed, only 68.47% of the reviews reported the FU definition, taken from the analysed case studies (see Sect. 2.1). As regards S-LCA, even if the FU is identified in 75.28% of the case studies (see Sect. 2.2), this definition does not go further than the FU definition in the UNEP/SETAC [7]. Hosseinijou et al. [13], Yıldız-Geyhan et al. [14], Raffiani et al. [15] highlight the difficulty of linking the FU to the LCIA phase, being data in S-LCA qualitative and semi-quantitative. Consequently, the social impacts are evaluated with regard to the behaviour of the company rather than to the input and output flows of processes.

1.7 Conclusions and Future Developments

This article represents a preliminary phase of investigation regarding the GSD phase of an LCT study. The FU definition is an important aspect of the LCA methodology when it comes to the modelling of the product system under analysis and thus it is a common practice. On the other hand, the FU in S-LCA does not seem to be easily identifiable. This article analysed the FU definition in case studies in both LCA and S-LCA via a literature review in order to detect its selection criteria.

The results of the study showed that the FU can be defined in a similar way for both methods in the various analysed sectors. This statement can therefore show that the FU selection depends on the product rather than on the orientation of the analysis (environmental or social). In addition, the results showed a prevalence of the use of mass as a quantity for FU identification in both methodologies. Considering that the FU should focus on the functional aspects, the prevalence of mass can be open to criticism. However, this can be justified, e.g., since it is one of the simplest quantities to be applied or because it is influenced by the choice of the reference flow. This aspect should be studied extensively, also in view of the differences between a stand-alone and a comparative analysis. Indeed, in the latter, the FU selection can strongly influence the results and, therefore, the selection of an "easy" FU is not always adequate. Furthermore, with a view to carrying out a sustainability assessment (Life Cycle Sustainability Assessment), it is necessary to combine the results of S-LCA and LCA and to identify a single FU. For these reasons, further developments of this work will include in the analysis the results of the various case studies (from the LCIA phase) with respect to the used FU. In this way, it will be possible to identify the way in which the results in both methodologies can be influenced by the choice of the FU. Therefore, it will be possible to acquire a complete picture of the dynamics of FU definition and application in the case studies of LCT.

References

1. Finkbeiner M, Schau EM, Lehmann A, Traverso M. Towards life cycle sustainability assessment. Sustainability. 2010;2(10):3309–22.
2. Kloepffer W. Life cycle sustainability assessment of products. Int J Life Cycle Assess. 2008;13 (2):89–95.
3. ISO 14040. Environmental management – life cycle assessment – principles and framework. Geneva, 2006.
4. D'Eusanio M, Zamagni A, Petti L. La social life cycle assessment a supporto del supply chain management, 11th conference of the Italian LCA network, resource efficiency e sustainable development goals: il ruolo del life cycle thinking, Siena, 2017, p. 279–287.
5. Macombe C, Feschet P, Garrabé M, Loeillet D. 2nd International seminar in social life cycle assessment – recent developments in assessing the social impacts of product life cycles. Int J Life Cycle Assess. 2011;16(9):940–3.
6. Di Cesare S, Silveri F, Sala S, Petti L. Positive impacts in social life cycle assessment: state of the art and the way forward. Int J Life Cycle Assess. 2016:1–16.
7. United Nations Environment Programme and Society for Environmental Toxicology and Chemistry, Guidelines for social life cycle assessment of products, Paris, 2009.
8. Weidema B, Wenzel H, Petersen C, Hansen K. The product, functional unit and reference flows in LCA, environmental news 70, 2004, Danish Ministry of the Environment – Environmental Protection Agency.
9. http://biblauda.unich.it/?152. Accessed 30 Oct 2017.
10. http://ec.europa.eu/competition/mergers/cases/index/nace_all.html. Accessed 10 Apr 2018.
11. Arzoumanidis I, Raggi A, Petti L. Environmental assessment of beekeeping products and services – a life cycle assessment case study including honey and pollination, proceedings of the 10th congress of the Hellenic Society of Agricultural Engineers, Athens, 2017, p. 426–435.
12. Bjørn A, Laurent A, Owsianiak M, Olsen SI, History LCA. In: Hauschild MZ, Rosenbaum RK, Olsen SI, editors. Life cycle assessment – theory and practice. Cham: Springer; 2018. p. 17–41.
13. Hosseinijou SA, Mansour S, Shirazi MA. Social life cycle assessment for material selection: a case study of building materials. Int J Life Cycle Assess. 2014;19(3):620–45.
14. Yıldız-Geyhan E, Altun-Çiftçioğlu GA, Neşet Kadırgan MA. Social life cycle assessment of different packaging waste collection system. Resour Conserv Recycl. 2017;124:1–12.
15. Raffiani P, Kuppens T, Van Deal M, Azadi H, Lebailly P, Van Passel S. Social sustainability assessments in the biobased economy: towards a systemic approach. Renew Sust Energ Rev. 2018;82(2):1839–53.

Chapter 2
Towards a Taxonomy for Social Impact Pathway Indicators

Bo P. Weidema

Abstract A conceptually complete taxonomy is proposed at three levels of the impact pathway: Elementary flows, midpoint impacts, and endpoint impacts. The completeness is ensured conceptually by including unspecified residuals and by the use of fully quantifiable indicators that can be traced from source to sink, so that completeness can be verified by input-output balances and against measured totals. Each category in the taxonomy has a definition and at the lowest level also a unit of measurement. Examples of category definitions and units are illustrated in an impact pathway model with starting point in the midpoint impact category "Undernutrition". This model also demonstrates the role of the taxonomy in the development of characterisation factors.

2.1 Introduction

The purpose of taxonomy is to provide structure and conceptual clarity to a scientific domain through clear definitions of hierarchically organised concepts. By reducing confusion and supporting harmonisation of terminology, the ultimate purpose is to improve monitoring, knowledge-generation, and decision-making. For social impact pathway indicators an important aspect of this is to ensure consistency in modelling, so that similar impacts are treated in a similar way.

Social impacts are here understood in the wider sense of welfare economics, as all impacts that affect human wellbeing, including ecosystem, health and socio-economic impacts.

The concept of impact pathway indicators has its own taxonomy, with the most well-known being the DPSIR framework of EEA [1], dividing indicators in Driving Force, Pressure, State, Impact, and Response indicators. Within the field of Life Cycle Assessment, as standardised in the ISO 14040 series, the same impact pathway indicators have different names as shown in Table 2.1 Here, the latter

---

B. P. Weidema (✉)
Danish Centre for Environmental Assessment, Aalborg University, Aalborg, Denmark
e-mail. bweidema@plan.aau.dk

© The Author(s) 2020
M. Traverso et al., *Perspectives on Social LCA*, SpringerBriefs in Environmental Science, https://doi.org/10.1007/978-3-030-01508-4_2

Table 2.1 Classes of impact pathway indicators in the EEA and LCA

DPSIR [1]	LCA (ISO 14040 series)
Driving force	Functional unit, Reference flow or Intermediate flow (between economic processes)
Pressure	Elementary flow
State	(no parallel, except when describing a baseline, reference, or background situation)
Impact	Impact category endpoint (often shortened to "impact" with indicators divided in *midpoint indicators* and *endpoint indicators*, the latter often classified in *Areas of Protection*)
Response	(no direct parallel; Responses may be formulated as new Functional Units of different improvement scenarios)

terminology is applied, except for the use of the term "pressure" in the example in Sect. 2.9.

Contributions towards a taxonomy for social impact pathway indicators have been made by:

• Jolliet et al. [2], in particular for Areas of Protection;
• Bare & Gloria [3], who presented a very detailed taxonomy, however limited to physical impacts and introducing a concept of "mode of contact" as a midpoint between elementary flows and midpoint impacts, although this did not play a central role in structuring their taxonomy;
• Simões [4], who collected 1450 social indicators from 51 documents from more than 30 scientific journals and classified these into 54 indicator families, further classified according to the 22 social aspects of the Global Reporting Initiative – a classification that is most relevant at the level of elementary flows, but which does not consider the further cause-effect relations required for linking to midpoint and endpoint indicators;
• UNECE [5] providing a very comprehensive set of sustainability indicators and a very clear description of the relationship between these indicators and the national accounting framework, particularly pointing out that for each aspect to be covered, both a geographical (imports/exports) and a temporal (transfer to future generations) perspective need to be covered.

The taxonomy presented here extends these contributions by suggesting a conceptually complete taxonomy at three levels of the impact pathway: Elementary flows, midpoint impacts, and endpoint impacts. The completeness is ensured conceptually by including unspecified residuals, but also and more importantly by the use of fully quantifiable indicators that can be traced from source to sink, so that completeness can be verified by input-output balances and against measured totals.

A distinction between biophysical, economic and social indicators has been maintained at the level of elementary flows, while for midpoint impacts the social and economic melts together as socio-economic indicators. When values are introduced at the level of endpoints (areas of protection), it is no longer meaningful to maintain the distinction between biophysical and socio-economic, even though some impacts can still be measured in physical units.

2.2 Equity-weighted Welfare ("Utility") as Single-score Endpoint

In accordance with welfare economics, the taxonomy applies equity-weighted welfare (or "Utility" for short) as single-score endpoint indicator. The equity-weighting (also known as utility-weighting, welfare-weighting, or distributional weighting) is necessary to take into account that the same impact is more burdensome (and that a similar improvement is more valuable) for individuals with lower income, and also allows a distinction between the weights given to impacts that directly affect wellbeing versus impacts that affect wellbeing indirectly via changes in productivity [6]. Thus, utility is measured in equity-weighted and purchasing-power-corrected monetary units. When communicating values, the most appropriate unit should be chosen, depending on the audience. The use of monetary units for communicating values should be limited to those situations where it is desired by the audience. Single-score results may, e.g., also be expressed in sustainability-points or Quality-Adjusted person-Life-Years. Monetary units are simply preferred for convenience by many decision-makers. The advantage of a single-score endpoint is that it allows explicit trade-offs to be made between the indicators of the different Areas of Protection. The inclusion of a single-score endpoint in the taxonomy does not imply that single-score methods have to be used in order to benefit from the remaining part of the taxonomy.

2.3 Areas of Protection

A conceptually complete organisation of "areas of protection" was suggested by the UNEP/SETAC Working Group on Impact Assessment [2]. Table 2.2 shows this with a few modifications. What is meant here by conceptually complete is that any item must be either human or non-human; any non-human item must be either biotic or non-biotic; any item must have either intrinsic value (be valuable in itself) or instrumental value (be valuable as a means to an end). What is here called "Instrumental" may also be called "Resources" or "Capital".

In the definition of the WHO [7], human health is "a state of complete physical, mental and social well-being and not merely the absence of disease or infirmity",

Table 2.2 Areas of protection in the SETAC/UNEP LCIA framework from [2], slightly modified by Weidema [8] by adding the terms in brackets

Objects considered → Endpoint value ↓	Humans	Biotic environment (natural and artificial)	Abiotic environment (natural or artificial)
Intrinsic	Human health (and well-being)	Biodiversity (and well-being of animals in human care)	Natural and cultural heritage
Instrumental	Human productivity	Ecosystem productivity	Natural resources and man-made capital

implying that the term also covers human wellbeing in a wider sense. However, in practice, the definition is used in the more narrow sense of mortality and morbidity as reflected in the use of DALY (Disability-Adjusted Life-Years) as a unit of measurement (e.g., in the Global Burden of Disease studies). DALY is also the typical unit used for the human health impact category indicator in most LCIA methods. Some models, especially those including social impact pathways, instead use the unit of QALY (Quality-Adjusted Life-Years), to reflect the wider wellbeing perspective.

The term "endpoint" for the indicators of the areas of protection implies that these are seen as independent and non-interacting. For the impact pathways, this implies that a midpoint impact that ultimately affects more than one endpoint should have an impact pathway to each of these endpoints. For example, a disease will typically both have a pathway to human health and a separate pathway to human productivity (lost workdays and health care costs). When a single-score endpoint is applied, the "areas of protection" endpoints effectively become midpoints towards the single-score endpoint. The use of the term "endpoint" is thus context-dependent.

2.4 Midpoint Impact Categories

Midpoint impacts can both affect endpoints and other midpoint impact categories. In Table 2.3, midpoint impact categories at the two top levels are listed. In the full taxonomy, a third level exists for many midpoint impact categories, and the taxonomy is open for further refinement. For example, the level 2 category "Inadequate maternity support" has a sub-category "Food insecurity" at level 3, defined as "Prevalence of insufficient amount and quality of individual food intake among children and women of childbearing age" and measurement unit: "Dimensionless ratio representing affected fraction of population (prevalence)". Further examples of definitions and units are provided in Sect. 2.9.

The majority of the midpoint impact categories in Table 2.3 are relatively self-explanatory. However, the one named "market distortion" is an aggregate of many more specific midpoint impact categories, and may therefore need to be explained here. It can be differentiated by the more specific nature of inequality of opportunity and transaction conditions (e.g., information inequality, discrimination, trade barriers) and by market (which includes markets for production factors). What is common for all of these is that different market actors are treated unequally or even completely prevented from access to a specific market.

2.5 Elementary Flow Categories

For the elementary flows, the top levels (see Table 2.4) are relatively aggregated, especially for the biophysical pressures, where level 3 (not shown in Table 2.4) contains 37 categories, and many more, e.g., specific substance emissions, at level

Table 2.3 Top-level midpoint impact categories

Level 1	Level 2
Biophysical impacts	Acidification
–	Antibiotic resistance
–	Aquatic oxygen depletion
–	Eutrophication
–	Global warming, ecosystem impact
–	Global warming, human impact
–	Human disease from respiratory particulates
–	Human toxicity
–	Other human diseases
–	Other physical impacts*
Socio-economic impacts	Absolute poverty
–	Capital market failure
–	Government failure
–	Human migration, forced
–	Inadequate access to pensions or social security
–	Inadequate maternity support
–	Inadequate conservation of cultural heritage
–	Inadequate social infrastructure*
–	Insufficient health care system
–	Insufficient skills
–	Market distortion, except capital markets
–	Underinvestment in education
–	Underinvestment in health care
–	Underinvestment in natural disaster damage prevention and mitigation
–	Underinvestment in physical infrastructure
–	Unemployment and underemployment
–	Unwanted pregnancy

*Unspecified residuals are indicated with an asterisk

4. For the economic and social pressures, level 3 categories are shown in Table 2.5. Each flow category has a definition and at the lowest level also a unit of measurement.

2.6   Modelling the Impact Pathways

Impact pathway modelling can take its starting point in an elementary flow, a midpoint or an endpoint, and thus model both forwards and/or backwards in the impact pathway. Modelling backwards in the direction of elementary flows ensures that the full impact can be allocated to its causes, and is thus recommendable. Causal relationships can best be expressed as marginal characterisation factors (unit of

Table 2.4 Top-level elementary flow categories

Level 1	Level 2
Biophysical pressures	Biological contamination
–	Direct physical changes to environment
–	Dissipative use of natural resources
–	Energy emissions
–	Overconsumption
–	Substance emissions
Economic pressures	Human time (labour & leisure hours)
–	Insufficient payment of labour or taxes
–	Monetary expenditure, except wages
Social pressures	Illegitimate resource acquisition and control
–	Inadequate work environment
–	Violence

endpoint indicator per unit of midpoint indicator or elementary flow indicator, unit of resulting midpoint indicator per unit of causing midpoint indicator or elementary flow indicator), allowing direct calculations of impacts by matrix inversion [9].

2.7 An Example of a Social Impact Pathway Model: Undernutrition

The principle of the impact pathway modelling is illustrated in Fig. 2.1 with a starting point in the midpoint "Undernutrition" (level 3 midpoint category under "Other human diseases") and its two further sub-categories: "Sub-optimal infant feeding practices" and "Childhood and maternal undernutrition". The extent of undernutrition is know from statistics, which allows a complete breakdown to causal factors, using on the one hand known cause-effect relationships and on the other hand a residual pathway. In the case of "Sub-optimal infant feeding practices" this residual pathway is "Insufficient health care system", and for "Childhood and maternal undernutrition" it is "Food insecurity", both having "Underpayment of labour or taxes" as the ultimate residual elementary flow.

2.8 Pressure Categories and Indicators (1–5) for Undernutrition

This Section provides definitions of the five pressure categories and indicators, in LCA terminology known as inventory indicators, that contribute to undernutrition. The first four occurs in productive activities, while the last (household gender

Table 2.5 Level 2 and 3 elementary flow categories for economic and social pressures

Level 2	Level 3
Human time (labour & leisure hours)	Labour hours
–	Leisure hours
Insufficient payment of labour or taxes	Extreme underpayment of labour
–	Underpayment of labour or taxes
Monetary expenditure, except wages	Net distortionary taxes
–	Net externality-correcting taxes
–	Net non-distortionary taxes
–	Net operating surplus
–	Rent
–	Voluntary financial transfers
Illegitimate resource acquisition and control	Burglary or attempted burglary
–	Illegitimate acquisition and control of physical resources
–	Rent seeking
–	Trafficking of humans
Inadequate work environment	Bonded labour
–	Child labour
–	Excessive work
–	Inadequate ergonomic condition
–	Insufficient paid breaks for breastfeeding
–	Premature return to work after giving birth
–	Stressful work condition
Violence	Genital mutilation
–	Incarceration
–	Infringement of freedom of expression
–	Interpersonal or communal violence
–	Participation restriction
–	Reduction in well-being of animals in human care
–	Refugees warehousing
–	Threat of violence or other contact crime
–	Threatening or traumatic traffic situations

discrimination) occurs in the sphere of private households and is not related to any product life cycles (in contrast to production activities).

Pressure Category (1): Insufficient Paid Breaks for Breastfeeding An important cause of undernutrition is *premature cessation of exclusive breastfeeding*, which is affected by insufficient breaks for breastfeeding at the workplace. Employers can guarantee paid breastfeeding breaks and thus reduce this cause of cessation of breastfeeding.

Pressure indicator: Number of annual female full-time employees without legal or contractual guarantee of a minimum of three daily paid breaks for breastfeeding

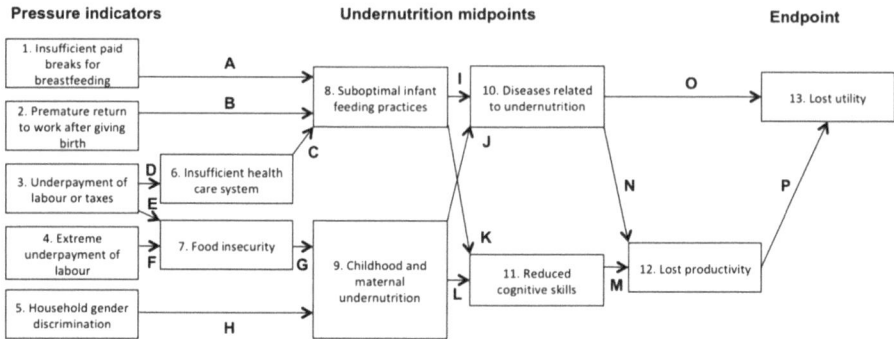

Fig. 2.1 Impact pathways for undernutrition in flow-chart format. Numbers and letters refer to sections in the text where each relation is described and quantified

providing sufficient time to express and deliver the breast milk to the child until the age of 1 year. Unit: (employment-) year or equivalent.

Pressure Category (2): Premature Return to Work After Giving Birth In a US cohort study of singletons whose biological mothers worked in the 12 months before delivery, Ogbuanu and co-workers [10] found that cessation of breastfeeding was not correlated to *length* of maternity leave (which does not need to be taken consecutively), but rather with first return to work. The indicator should therefore reflect requirements for early return to work, rather than the length of the maternity leave.

Pressure indicator: Number of annual female full-time employee equivalents without legal or contractual guarantee of a continuous period of maternity leave until the child has an age of 6 months. Unit: (employment-) year or equivalent.

Pressure Category (3): Underpayment of Labour or Taxes An important cause of undernutrition is poverty, inequality, and insufficient governance, leading among other things to food insecurity and insufficient health care systems (the latter partly via an intermediate midpoint impact category *Underinvestment in health care* that is not shown in Fig. 2.1). At a very general level, all of these impacts can be related back to insufficient funding, either directly through insufficient wages or directly or indirectly through insufficient income for managing public services. The pressure category "Underpayment of labour or taxes" is thus a very generic category that acts as a default starting point for all impact pathways that cannot (currently or by their nature) be related to more specific pressure indicators.

Pressure indicator: The difference between the current World-Bank-purchasing-power-corrected labour and tax expenditures and the labour and tax expenditures for the same amount of work hours in an ideal situation without avoidable social externalities, as defined by Weidema [6]. Unit: Purchasing-power-corrected currency units (e.g. $USD_{2017,PPP}$).

Fig. 2.2 Simplified relationship between income level and prevalence of malnutrition, based on data for Pakistan [11], updated to 2011 income levels

Pressure Category (4): Extreme Underpayment of Labour Extreme underpayment of labour is the form of underpayment that leads to extreme absolute poverty (as opposed to relative poverty) where the ability to purchase essential goods is affected. The relationship between income and malnutrition, see Fig. 2.2 in Sect. 7.2, indicates that 6 USD_{2011}/day/person is the poverty line below which malnutrition begins to occur, and that a sharp increase appears at 4 USD_{2011}/day/person. Since the average amount of labour hours per day per person in 2011 is 3.87 (27 hours per week, year-round, implying that each person in full-time work provide on average for slightly less than one person out of work), the two thresholds are met for wage exceeding 1.55 and 1.03 USD_{2011}/work-hour, respectively. The underpayment is the difference between the actual payment and these poverty lines.

Pressure indicators: Accumulated differential between the World-Bank-purchasing-power-corrected labour expenditures and the poverty line of 1.55 $USD_{2011,PPP}$/ work-hour, subdivided in the upper level between 1.55 and 1.03 $USD_{2011,PPP}$/work-hour and the very extreme underpayment below 1.03 $USD_{2011,PPP}$/work-hour. Unit: Purchasing-power-corrected currency units.

Pressure Category (5): Household Gender Discrimination Household gender discrimination is a level 4 pressure category under *Illegitimate acquisition and control of physical resources*. It can lead to *Childhood and maternal undernutrition* both in the presence of food insecurity at the household level and in households that on average are deemed to be food secure, when distribution of food within the household is skewed in favour of male household members, and indirectly through adolescent maternity and maternal depression.

Currently, the only generally available proximate indicator of household gender discrimination is that of intimate partner violence (IPV). Furthermore, IPV can in itself be seen as an actual cause for *Childhood and maternal undernutrition*, since women are more likely to have a stunted (undernourished) child if they have experienced physical intimate partner violence. This is supported by evidence of a pathway from IPV through adolescent maternity and maternal depression, both influencing nutritional status of mother and child.

Pressure indicator: Number of women with lifetime experience of physical violence. Alternative pressure indicator: Number of women with experience of physical violence within the last year. Unit: Persons.

2.9 Midpoint Impact Categories and Indicators (6–11)

Midpoint Impact Category (6): Insufficient Health Care System This impact category captures all avoidable causes of disease. This implies a rather broad definition of "health care system" to include also – and maybe in particular – preventive activities. It is estimated that the impact of the health care system on undernutrition is primarily related to the (insufficient) advice given to mothers. No separate outcome indicator is suggested for this advice, which implies that the same outcome indicator is used given under for the subsequent impact category (8): *Suboptimal infant feeding practices.*

Midpoint Impact Category (7): Food Insecurity The overall effect of underpayment via food insecurity to childhood and maternal undernutrition is modelled by the direct income poverty relationship given by Blakely and co-workers [11] and shown in Fig. 2.2, which provides a direct relationship to the pressure indicator of *Extreme underpayment of labour.* Any food insecurity not captured by this direct relation will be captured by the indirect pathway from *Underpayment of labour or taxes,* covering insufficient income redistribution and insufficient funds for infrastructure.

Since the concern of the subsequent midpoint indicator *Childhood and maternal undernutrition* is limited to undernutrition for children and women of childbearing age, it is also this group that is particularly relevant to consider for the food insecurity indicator.

Midpoint indicator: Prevalence of insufficient amount and quality of individual food intake among children and women of childbearing age. Unit: Dimensionless ratio representing affected fraction of population (prevalence).

Midpoint Impact Category (8): Suboptimal Infant Feeding Practices (Subcategory of Insufficient Maternity Support) Undernutrition in infants and young children can be caused by poor feeding practices, especially insufficient breastfeeding and lack of responsive and timely complementary feeding, where the caregiver is responsive to the child clues for hunger and encourages the child to eat other foods than breast milk from the age of 6 months. To include duration of exclusive breastfeeding, the characterisation factor has been expressed in time units rather than per infant. However, this implicitly assumes that there is a linear relation between duration and impact within each assessed period (e.g., 0–6 months; 6–12 months).

Category indicators: Premature cessation of predominant breastfeeding earlier than 6 months after childbirth and *Discontinued breastfeeding earlier than 12 months after childbirth. Unit: Weeks, or equivalent time unit.*

Midpoint Impact Category (9): Childhood and Maternal Undernutrition (Subcategory of Other Human Diseases) This midpoint impact category covers both Protein-Energy-Undernutrition (PEU) and micronutrient deficiencies, which occur together, while micronutrient deficiencies may also occur separately. However, current data does not allow separate impact pathway descriptions for these two forms of undernutrition. Stunting (low height-for-weight) is the most appropriate measure for long-term, chronic undernutrition from the interaction of poor diet and repeated infections, often persisting even in situations of decreasing prevalence of wasting (low weight-for-height), which rather measures acute undernutrition, and underweight, which is a composite measure of both chronic and acute undernutrition.

Category indicator: Prevalence of stunting in children age 5 years and under (height for age two or more standard deviations below the median of the reference population according to the WHO Child Growth Standards). Unit: Dimensionless ratio representing affected fraction of population (prevalence).

Midpoint Impact Category (10): Diseases Related to Undernutrition The Global Burden of Disease Collaborative Network [12] provides annual country-specific aggregate measures in Years-of-Life-Lost, Years-Lived-with-Disease (summed in Disability-Adjusted-Life-Years, DALY) for diarrheal diseases, lower respiratory infections, measles and protein-energy malnutrition related to suboptimal breastfeeding and childhood undernutrition, as well as diseases related to deficiency in Iron, Vitamin A, and zinc.

Category indicator: Incidences of specific diseases attributable to undernutrition. As human health endpoint, this may be aggregated as Disability-Adjusted person Life-Years (DALY). Unit: Number of incidences of each disease. Can be aggregated in DALY for purposes of comparison.

Midpoint Impact Category (11): Reduced Cognitive Skills Cognitive skills are generally measured by standardised tests involving, e.g., multiple choice, sentence completion, short answer, or true-false. The outcomes are normalised to a mean of 100 and a standard deviation of 15 IQ points for the population in question. In the tradition from Lynn [13] the British mean of 100 is used as a global reference level for comparisons across populations.

Category indicator: Change in intelligence quotient. Unit: IQ points.

2.10 Endpoint Impact Categories (12–13) for Undernutrition

Area of Protection Indicator (12): Lost Human Productivity Human productivity is measured in Productivity-Adjusted person-Life-years (PALY), thus accounting for incidence and duration of the impact in person-Life-Years (LY), modified by a dimensionless impact severity factor between 0 and 1 for the relative change in production output (PA) of the affected population.

Category indicator: Relative change in production output per person-year. Unit: Productivity-Adjusted person-Life-years (PALY).

Single-score Impact Category and Indicator (13): Lost Utility *Category indicator: Utility (equity-weighted welfare). Unit: Purchasing-power-corrected and equity-weighted currency units (with indication of base year).*

2.11 Characterisation Factors

Referring to the letters in Fig. 2.1, characterisation factors can be provided for the different relationships between pressure indicators, midpoint indicators, and endpoint indicators. An example of a characterisation factor (A) relating premature cessation of breastfeeding (indicator 8), measured in weeks, to workplace pressure indicator (1) can be based on the results of the global study on paid breastfeeding breaks by Heymann [14], indicating that a guarantee of such breaks would increase the average rate of breastfeeding by 10%, translated into a duration of 5 weeks per child when including continued breastfeeding until 12 months of age. Local characterisation factors per female work-year can be obtained by combining this with the local annual birth rate (children/1000 persons) and the local inverse female labour participation rate (1/(female work-years/1000 persons). By using global averages for these factors, a global default value of *0.38 weeks of additional breastfeeding/female work-year with legal or contractual guarantee of paid breaks for breastfeeding* is obtained.

References

1. EEA. Environmental indicators: Typology and overview, Technical report No 25, Copenhagen, European Environmental Agency, 1999.
2. Jolliet O, Brent A, Goedkoop M, Itsubo N, Mueller-Wenk R, Peña C, Schenk R, Stewart M, Weidema BP. Final report of the LCIA definition study, Paris, Life cycle impact assessment programme of the UNEP/SETAC life cycle initiative, United Nations Environmental Programme, 2009.

3. Bare JC, Gloria TP. Environmental impact assessment taxonomy providing comprehensive coverage of midpoints, endpoints, damages, and areas of protection. J Clean Prod. 2008;16:1021–35.
4. Simões MGFP. Social key performance indicators – Assessment in supply chains, Master Thesis, Instituto Superior Técnico, Lisboa, 2014.
5. UNECE. Conference of European Statisticians recommendations on measuring sustainable development. New York and Geneva: United Nations; 2014.. www.unece.org/publications/ces_sust_development.html
6. Weidema BP. The social footprint – A practical approach to comprehensive and consistent social LCA. Int J Life Cycle Assess. 2018;23(3):700–9.
7. WHO. Preamble to the Constitution of the World Health Organization as adopted by the International Health Conference, New York, July 1946, Official Records of the World Health Organization, No. 2, p. 100.
8. Weidema BP. The integration of economic and social aspects in life cycle impact assessment. Int J Life Cycle Assess. 2006;11(1):89–96.
9. Weidema BP, Schmidt J, Fantke P, Pauliuk S. On the boundary between economy and environment in LCA, Int J Life Cycle Assess, early on-line view 4. October 2017, Read-only link: http://rdcu.be/wswU.
10. Ogbuanu C, Glover S, Probst J, Liu J, Hussey J. The effect of maternity leave length and time of return to work on breastfeeding. Pediatrics. 2011;127(6):e1414–27. https://doi.org/10.1542/peds.2010-0459.
11. Blakely T, Hales S, Woodward A. Poverty: assessing the distribution of health risks by socioeconomic position at national and local levels. Geneva: World Health Organization, WHO Environmental Burden of Disease Series, No 10, 2004.
12. Global Burden of Disease Collaborative Network, Global Burden of Disease Study 2016 (GBD 2016). Burden by Risk 1990–2016. Seattle: Institute for Health Metrics and Evaluation; 2017.
13. Lynn R, Meisenberg G. National IQs calculated and validated for 108 nations. Intelligence. 2010;38(4):353–60.
14. Heymann J, Raub A, Earle A. Breastfeeding policy: a globally comparative analysis. Bull World Health Organ. 2013;91:398–406.

Chapter 3
A New Scheme for the Evaluation of Socio-Economic Performance of Organizations: A Well-Being Indicator Approach

Silvia Di Cesare, Alfredo Cartone, and Luigia Petti

Abstract In this paper we propose to evaluate socio-economic performance of organizations through a well-being approach. Our aim is to build a composite indicator for product socio-economic impacts. As composite indicators are useful to simplify the behaviour of complex phenomena, a methodology based on well-being indicators is developed in the scope of the affected population. The organization actions are connected to the weights of the well-being indicators based on the effective links existing between these actions and the well-being dimensions. Thereafter, the links between variables from social reporting and life cycle inventory indicators are defined by conducting a Delphi expert consensus method on the basis of the "Wisdom of crowds" theory.

3.1 Introduction

The ultimate goal of sustainable development is human well-being, contributing to the needs of current and future generations [1].

In the field of product and process assessment, some methodologies, techniques and tools have been developed, mostly supporting policies and strategies for the social, economic and environmental dimension of sustainable development. In the language of economists, these tools are aimed to assess internalities and externalities of products/services along their entire life cycle [1]. One usual way of interpreting sustainability is to call upon three pillars. In this particular view, the economic pillar of sustainability is expected to be evaluated through the Life Cycle Costing (LCC) methodology. The environmental one, instead, is covered by the most used tool: environmental LCA (E-LCA). Its practitioners evaluate the impacts of product life

S. Di Cesare (✉)
Department of Economic Studies, University "G. d'Annunzio", Pescara, Italy

CIRAD, UPR GECO, Montpellier Cedex 5, France
e-mail: silvia.dicesare@unich.it; silvia.di_cesare@cirad.fr

A. Cartone · L. Petti
Department of Economic Studies, University "G. d'Annunzio", Pescara, Italy

M. Traverso et al., *Perspectives on Social LCA*, SpringerBriefs in Environmental Science, https://doi.org/10.1007/978-3-030-01508-4_3

cycles according to "Areas of Protection" (AoP). These are "domains" that need to be preserved and indicate the impact categories of value to society. There is consensus on the nature of the AoP in E-LCA (human health, natural resources, natural and man-made environments).

Clear measurement of the social performance of an organization is still a lively field of research and so it is the definition of a valid methodology to work out the social impact of a single product or service. Existing Social Life Cycle Assessment (S-LCA) case studies do not actually evaluate the social performance of products. From the 189 indicators proposed in S-LCA, only eight refer to the product level, while 127 refer to the organizational level and 69 to country level—including overlaps and according to the methodological sheets [2]. This circumstance clearly leads to an organizational approach of S-LCA named Socio-Organizational Life Cycle Assessment (SOLCA) [3] based on the Organizational-LCA (O-LCA) model [4]. While not directly referring to the O-LCA methodology, this work intends to address this type of approach to assess social impacts.

In general, a clever way to assess a complex reality is by building a composite indicator able to include a multivariate reality into a single number. Composite indicators are useful to simplify the essence of complex phenomena and for this reason are extensively adopted. However, great attention must be given to the definition of a composite indicator in order for it not to be misleading. In the field of LCA, the use of single and composite indicators is largely diffused as they are involved in the evaluation of and in many of the steps that lead to a final assessment. Specifically, in S-LCA, a variety of scientifically recognized methodologies can be implemented towards the aim of synthetizing the large number of data.

Following [5], the existing SLCIA methods can be classified into two broad categories: type I (performance reference point) and type II (impact pathways methods). These categories can be further divided into subcategories they are checklist method, scoring method, Social Hotspot DataBase (SHDB) method for performance reference point [6]. Identify one of the main approaches in type I characterization as that "based on stakeholders' or experts' judgment of companies'/sectors' compliance to societal expectations or norms". In [7] stakeholders are asked to assess, on the basis of their perception, the level of compliance with social compliance criteria by companies/organizations within a recycling system.

However, in S-LCA, few methodologies consider appropriately the extensive importance of well-being as a multivariate phenomenon that recollect several aspects of human life. Evaluating the consequences of organization behaviour on well-being is vital to elaborate a valid model to assess social performance. In fact, well-being should, to some extent, represent the basis for evaluating an organization in terms of social performance. In this sense, one of the major supply of an organization could be the contribution to the improvement of the society that can be measured in term of widespread amelioration of stakeholders' well-being. Hence, rapid changes in worldwide economy call us to take into account a wide concept of sustainability which includes social sustainability [8].

If this approach is largely synthetized into the conceptual frame of E-LCA particularly in the Area of Protection (AoP) "Human Health" (HH), a vast literature has recognized the importance of outperforming an assessment procedure based on DALY (Disability Adjusted Life Years).

Although the concept of DALY has proven to be a useful metric in the assessment of human health damage in LCA, it has been criticized due to some methodological aspects considered as subjective:

1. DALY refer to a specified region and time frame, such as the world in 1990 [9]. Thus, applying world average DALY estimates in the calculation of characterisation factors implies acceptance of the assumption that damage to human health due to life cycle emissions can be represented by world averages. However, for LCA case studies focusing on region-specific human health impacts, DALY estimates should be carefully considered. In fact, taking another region in the world as a starting point for the DALY calculation, may cause a change in the results [10].
2. Secondly, in most LCIA methodologies, DALY is calculated without applying age-specific weighting and without discounting future health damages. These two assumptions, however, are disputable.
3. Thirdly, the use of YLD (Years Lost due to Disability) includes a subjective assessment of the weighting of health disabilities [11]. The difficulties linked with such an assessment explain why some of the LCIA methodologies explicitly exclude YLD from the damage assessment.

In E-LCA the approach based on DALY represents a well-known and concise method of assessment to obtain a synthetic measure of the organizational performance. Nevertheless, to the best of our knowledge, there are difficulties in S-LCA applications to lead the practitioner to a complete and final evaluation of the social performance. Hence, a complete evaluation of social performance of an organization is still difficult due to several reasons. This can occur because of the lack of data or of the lack of a precise set of instruments which takes into account the impact on dimensions which can measure an increase of well-being.

A common and structured approach to assess social impacts trough a consensus method cannot be found in S-LCA applications [12], and a proper evaluation of the different techniques applied is difficult since only a few cases describe questionnaires, groups, and number of people involved in a detailed way [13]. Thus, the scoring process in indicators of S-LCA is widely based on the mere consensus of a selected panel of experts or on the practitioner experience. This aspect is likely to introduce a wide discrepancy between the reality and the evaluation process due to the subjectivity of the assessment.

In this last decade, a vast literature has emphasized the importance of a more accurate and reliable set of indicators to summarise a variety of economic performance, e.g. deprivation [14] and well-being [15]. This literature broadens the representation of the economic performance of countries or regions or cities beyond the exclusive focus on the GDP as the only measure of development [16]. The representation of the economic reality as a complex phenomenon is also stated in the well-known specification of the three pillars of sustainability.

Specifically, well-being indicators picture this multi-dimensionality by trying to consider all the domains which influence the conditions of people and communities. This result is obtained by providing accurate data and reliable sets of weights, which enable the synthesis of a large number of variables into a single number. The building of these weights is a crucial point in the field of economic statistics and the weighting schemes can be derived in different ways [17]. Nevertheless, a critique of the different approaches to build a composite indicator of well-being is out of the scope of this work. Thence, in this paper the aim is to rely on the literature of well-being indicators to face the problem of scoring using a novel approach.

Starting from a well-being indicator a definition of the latter is obtained and allows for a measurement of the impact of organization actions. The approach, based on an equal evaluation of different capitals, and not only on the economic one, reconnects to the Capacities S-LCA approach [18], which is based on the Multi Capital Model (MCM) rooted in Sen's Theory of Capabilities [19]. Thus, according to SOLCA methodology the evaluation is carried out taking into account the organization which allows for higher flexibility and a potentially context related approach.

3.2 Method

Tipically, in S-LCA two possible ways to carry out Impact Assessment (IA) are presented. These have been called Type I and Type II. Type I, or social life cycle attributes assessment (S-LCAA) [20, 21] does not provide a quantitative measurement of social impacts for two reasons: it is in the sphere of the only internal corporate performance, and, therefore, offers the point of view of the producer of social actions; it depicts a static situation (so can't account for the impacts stemming from change). On the other side, Type II, or "pathways" analysis, looks for statistically significant relations between factors and impacts [22, 23]. It has firstly been implemented by [21] in the second part of his paper, and [24] determining social impacts on human health resulting from a change in products' life cycles [25]. Within this type II, [18] specify an approach called "Capacities social LCA", which is rooted in the Sen's theory of Capabilities.

Focusing on the Capacities S-LCA approach, its principle is to articulate a chain analysis with an MCM approach retaining five classes of capital (human, natural, institutional, social, economic capital), in order to measure the variations of capacities of the actors, resulting from the social practices of organizations. The point is not to measure a behavioural performance of social responsibility, but to measure an impact on the actual potential capacities and even on the real capacities of the actors. The proposed methodology could be placed in the field of Capacities S-LCA because its aim is to assess in which way organizational behaviour could impact on the different capitals considered as an input in the MCM approach.

This methodology individuates a well-being indicator in the scope of the affected population. Then, we connect the organization actions to the variables of the well-being indicators based on the effective links existing between these dimensions of well-being and the actions.

The links would be individuated by conducting a Delphi expert consensus method [26]. The validity of the approach would be supported by the theory of "wisdom of crowds" [27] showing that groups can make good judgements under certain conditions. This theory affirms that by aggregating many imperfect estimates, the group could make a much better estimate than the most skilled individuals. For this reason, several experts need to be elicitated in the field of organization practices and context experts. The experts elicited could vary from different disciplines implicated in organization management (e.g. experts in industry management sciences) and policy makers in different fields. In fact, it is possible to develop a consensus model of expertise through an iterative process of individual elicitation on a set of elements, assembly of the results and re-elicitation on the new set of elements [28].

The consensus changes over time as knowledge increases. For this reason, it would be advisable to associate a consensus and a Delphi method. The Delphi method is one of many that have been used to build expert consensus. Sometimes consensus builds rapidly and spontaneously in science, based on a critical piece of evidence. In this special case, this method enables to individuate solid connections between the operational of the organization and the variables that compose human well-being. In this scope, unpredictability, incomplete control, and plurality of legitimate perspectives have to be faced [29]. In such a context, a resort is claiming for expert elicitation. Indeed, the idea is that the expert experiences encompass (and can stand for) all the complex system of relationships embedded in the issue. Thus, solid connections could be developed by expert experiences. In this case, expert consensus would be based purely on personal experience, so this is the case that can be called "practice-based evidence".

In the suggested methodology, experts would be asked to set connections between inventory indicators and variables identified as dimensions of well-being. In fact, experts would be invited to converge around the potential effect of an organization action on the well-being spheres on one or more well-being dimensions included into a composite indicator. In practice, experts could choose between values 0 and 1 depending on the existence of an effect of an organization action on each variable of the composite indicator.

This paper contributes to the existing literature in two ways. Firstly, we use a consensus method to depict a variety of links between the dimensions of well-being and key indicators referring to the social indicators of the organization. This is crucial to develop a frame in which the actions of the organization can be synthetized to develop concise measures of social performances. The second contribute is a methodological development in the field of S-LCA IA methods in line with recommendation of [1]. Specifically, we design a scheme of evaluation based on a comprehensive approach that allows us to produce more precise and accurate scoring processes on the base of well-being indicators weights.

3.3   Results

A composite indicator of well-being may be expressed as sum of weighted values from a dataset including a wide range of variables. Variables may change according to the definition of the multivariate phenomenon. In general terms, given a set of P variables X collected for measurement of well-being for N units, a composite indicator of well-being may be synthetized for each unit as:

$$WB_i = \sum_{p=1}^{P} x_{ip} l_p \tag{3.1}$$

The weights l_p could be measured according to different techniques for which a review is in [30]. Additionally, those weights represent the relevance that the selected dimensions of well-being, identified as broad class of indicators, have for the class of stakeholders under focus. The problem of different scales among variables is usually solved according to normalization and standardization techniques.

The single indicators (i.e. dimensions) used for measuring well-being in a selected area are usually logically connected to a wide class of actions that can be implemented by an organization. For this reason, it is straightforward to think that the judgement that stakeholders gives to a single action would fall into the area of one or more single indicators of well-being. Thence, the intuition is to adopt weights from well-being indicator calculated for a class of stakeholders across an area for scoring actions of the studied organization. This could offer a new scheme to assess organization actions according to the preference that the stakeholders give to this particular dimension of well-being.

Consequently, it could be assumed that the social performance for an s category of stakeholder inside an organisation could be represented as the weighted sum reported in Eq. 3.2. Given a large number of J inventory indicators, selected as quantitative variables normalized in a unit measure (e.g. monetary values), we have that:

$$SP_S = \sum_j \sum_p w_{jp} z_j l_p \tag{3.2}$$

Where:

- z_j are values from a set of quantitative variables, i.e. inventory indicators or accounting data needed to evaluate the organization's performance and expressed in a unit measure.
- l_p are the weights used to synthetize the operational of the firm according to the relevance that the stakeholders give to each dimension of well-being (economic, social, environment, health, etc.).
- w_{jp} the degree of the effective relation between the action performed by the organization and what expected in that field by the stakeholders. Values w_{jp}

from a binary matrix W may assume values 0 or 1 based on the real presence of a link between the inventory indicators and each dimension of well-being considered. The matrix of binary values W has dimension JxP and is defined based on the results of a Delphi Expert Consensus Methods [26].

In Eq. 3.3 the overall social performance indicator based on the contribution of the organization on well-being is described.

$$SP = \sum_S SP_S \tag{3.3}$$

Note that in that specification each stakeholder has the same relevance.

Operatively, a first approach to build indicators could be to sum appropriate indicators without any sort of weighting. Alternatively, a well-known technique to build a set of weights is Principal Component Analysis (PCA) [31]. PCA decomposes the covariance structure into eigen vectors and eigen values. The eigen vectors (i.e. loadings) are used as weights for the data and synthesis of the variables. Therefore, a possible way to derive weights in well-being indicators is to rely on the structure of the phenomenon got by PCA. Frequently, results of the well-being indicator could be sensitive to the choice of the weighting scheme adopted. In fact, weights for the well-being indicator may be derived following a statistical technique, normative weights, or mixed approaches. Statistical techniques as PCA or regression models are considered an objective alternative as the weights are derived directly from data. Conversely, normative weights give to different relevance to dimensions, according to theoretical assumptions.

PCA, for example, could offer some drawbacks, particularly for two aspects. Firstly, PCA adopts a compensatory approach based on variance-covariance matrix spectral decomposition. Secondly, it discards the hypothesis that weights may change from a context to another [32, 33]. As the first problem could be successfully addressed by adopting not compensatory approaches based on statistical techniques [17], the second question remains open and widely interesting in order to set preferences schemes towards weights that are context related.

Furthermore, the hypothesis that the behaviour of the organization may be assessed in terms of well-being appears both theoretically plausible and effectively desirable. Particularly, for the problem of a correct estimation of the social performance, it represents a shortcut for the problem of the scoring. In fact, quantitative scoring is determined by the value assigned in accordance with beliefs about how something should be done. Quantitative scores are desirable as they offer easy to digest data on social impact. However, such scoring can also be reductionist, cloud transparency, and accentuate the subjectivity of measuring impact and at their worst, change behaviour to maximise scores, but possibly lessen overall more holistic economic impact. By offering a score, the user of the impact assessment has an indication of the performance of the company in that arena. Furthermore, for approaches such as the Social Return on Investment (SROI) framework, there is an attribution of monetary values to outcomes, which necessarily involves subjective judgement calls, especially when the outcomes have more social or political, rather than financial, implications attached.

3.4 Discussion

In economic analysis three main theories about well-being exist [23]. Well-being could be defined as the satisfaction of preferences [16], as happiness or satisfaction felt, and finally as conceived by the capabilities approach, developed by [19]. Following the latter, organization actions could be evaluated under the frame of the economy of capabilities [19]. based on their effort to tackle inequalities in a more general sense than pure economic inequality [23].

The principal goal of S-LCA fall into five categories that could be combined to the purposed approach and in the frame of the economy of capabilities. Providing knowledge about likely consequences of organization actions (e.g., what are the likely main impacts in terms of public health and in terms of workers' health), the use of the presented methodology could allow also policy makers in adopting strategies that could improve the level of multi-dimensional capital. Hence, what is highly relevant in S-LCA is helping coordination of actors involved outside and inside the organization (for instance, as a basis for discussions of the configuration of a project). The first step of this methodology is to consider an indicator of well-being, the following step is a detailed connection of all the quantitative indicators supplied by the management to the different variables that compose the MCM approach. Hence, influencing decision about future projects is important and S-LCA could be adopted as a decision support tool for evaluating future policies or actions. In this sense, the proposed approach could be used to evaluate broader strategies and policies still to implement. The studies stemming from S-LCA highlight the main social issues and claims for changes in the present project which may be marginal from the technical point of view, but very important from the social one. Moreover, S-LCA is called to help to fine-tuning the social side of projects. S-LCA fills in the social side of projects, by reporting on several social aspects (expected and unexpected), and by claiming for modifications when necessary. Here, by considering the well-being as a base for measuring impact in S-LCA a new perspective could be opened. By going beyond the idea of capital as a mere financial asset, a multidimensional concept of capital could represent the first step for recovering social aspects that could be marginal only at a first glance. Lastly, this technique is useful to generate innovations driven by social considerations and the main aim of this proposal is to offer a tool that highlights socially preferable alternatives. This that could favour an increase of several capital dimension including the human one.

3.5 Conclusion

Social dimension of sustainability should consider well-being as its main objective. The use of a multidimensional approach to well-being leads to consider the economy of capabilities as a compass that guides to a comprehensive and lean frame to assess social sustainability. In fact, to improve sustainability it is vital to reduce the level of inequalities and increasing capabilities of individual. In the presented methodology a scheme for evaluating the impacts on stakeholders' well-being is offered and extended into the field of S-LCA and SOLCA. Hence, positive and negative impacts are considered to assess the organization actions and indicators could be derived as context related. For the sake of simplicity, the proposed scheme was implemented for quantitative variables only, however a generalization to include both qualitative and quantitative data is possible.

As reality tends to be increasingly oriented towards multidimensional phenomena, the building of composite indicators will turn to be more and more meaningful. In the case of S-LCA and SOLCA, this means trying to build tools for measurement which offer direct connections of economy with social dimension of sustainability. Therefore, an achievement could be represented by the exposed contribute.

References

1. United Nations Environment Programme and Society for Environmental Toxicology and Chemistry. Guidelines for Social life cycle assessment of products. Paris; 2009.
2. UNEP/SETAC. The methodological sheets for sub-categories in social life cycle assessment (S-LCA). Life cycle initiative, UNEP-SETAC.
3. Martínez-Blanco J, Lehmann A, Chang YJ, Finkbeiner M. Social organizational LCA (SOLCA)—a new approach for implementing social LCA. Int J Life Cycle Assess. 2015;20 (11):1586–99.
4. UNEP/SETAC. Guidance on organizational life cycle assessment, 2015.
5. Chhipi-Shrestha GK, Hewage K, Sadiq R. Socializing'sustainability: a critical review on current development status of social life cycle impact assessment method. Clean Techn Environ Policy. 2015;17(3):579–96.
6. Garrido SR, Parent J, Beaulieu L, Revéret JP. A literature review of type I S-LCA—making the logic underlying methodological choices explicit. Int J Life Cycle Assess. 2018;23(3):432–44.
7. Arpacana S, Salhofer S. Application of a methodology for the social life cycle assessment of recycling systems in low income countries: three Peruvian case studies. Int J Life Cycle Assess. 2013;18(5):1116–28.
8. Boons F, Lüdeke-Freund F. Business models for sustainable innovation: state-of-the-art and steps towards a research agenda. J Clean Prod. 2013;45:9–19.
9. Murray CJL, Lopez AD. Evidence-based health policy--lessons from the global burden of disease study. Science. 1996;274(5288):740–3.
10. Goedkooop M, Heijungs R, Huijbregts M, De Schryver A, Struijs J, van Zelm R. ReCiPe 2008 A life cycle impact assessment method which comprises harmonised category indicators at the midpoint and the endpoint level, Report I: Characterisation, First edition (version 1.08).

11. Krewitt W, Pennington DW, Olsen SI, Crettaz P, Jolliet O. Indicators for human toxicity in life cycle impact assessment. Position paper for SETAC-Europe WIA2 task group on human toxicity, Version: January 7, 2002.
12. Mather S. Integrating participatory approaches into social life cycle assessment: the S-LCA participatory approach. Int J Life Cycle Assess. 2014;19(8):1506–14.
13. Foolmaun RK, Ramjeeawon T. Comparative life cycle assessment and social life cycle assessment of used polyethylene terephthalate (PET) bottles in Mauritius. Int J Life Cycle Assess. 2013;18(1):155–71.
14. Pampalon R, Raymond G. A deprivation index for health and welfare planning in Quebec. Chron Dis Inj Can. 2000;21(3):104.
15. ISTAT. Bes 2015: Il benessere equo e sostenibile in Italia, Istituto Nazionale di Statistica, 2015.
16. Stiglitz JE, Sen AK, Fitoussi JP. Rapport de la Commission sur la mesure des performances économiques et du progrès social, 2009.
17. Mazziotta M, Pareto A. A non-compensatory composite index for measuring well-being over time, cogito. Multidisciplinary research journal, Vol. 5, No. 4, 2013, p. 93–104.
18. Garrabè M, Feschet P. Un cas particulier: l'ACV sociale des capacités. ACV sociales. Effets socio-économiques des chaines de valeurs, FruiTrop Thema, Montpellier (France), 2013, p. 87–118.
19. Sen A. Social choice theory: a re-examination. Econometrica. 1977;45:53–89.
20. Andrews E, Lesage P, Benoit C, Parent J, Norris G, Revéret JP. Life cycle attribute assessment. J Ind Ecol. 2009;13(4):565–78.
21. Norris GA. Social impacts in product life cycles-towards life cycle attribute assessment. Int J Life Cycle Assess. 2006;11(1):97–104.
22. Macombe C. Researcher school book: social evaluation of the life cycle, application to the agriculture and Agri-food sectors. 2017.
23. Macombe C, Lagarde V, Falque A, Social LCAs. Socioeconomic effects in value chains, 1srt Editi. ed. FruiTrop, CIRAD. 2013.
24. Hutchins MJ, Sutherland JW. An exploration of measures of social sustainability and their application to supply chain decisions. J Clean Prod. 2008;16(15):1688–98.
25. Neugebauer S. Enhancing life cycle sustainability assessment, doctoral thesis. 2016.
26. Jorm AF. Using the Delphi expert consensus method in mental health research. Aust N Z J Psychiatr. 2015;49(10):887–97.
27. Surowiecki J. The wisdom of crowds. New York: Anchor, cop; 2004. p. 2004.
28. Léger B, Naud O. Experimenting statecharts for multiple experts knowledge elicitation in agriculture. Expert Syst Appl. 2009;36(8):11296–303.
29. Funtowicz SO, Ravetz JR. Uncertainty, complexity and post-normal science. Environ Toxicol Chem: Int J. 1994;13(12):1881–5.
30. Decancq K, Lugo MA. Weights in multidimensional indices of wellbeing: an overview. Econ Rev. 2013;32(1):7–34.
31. Jolliffe IT. Principal component analysis and factor analysis, principal component analysis; 2002. p. 150–66.
32. Cartone A, Postiglione P. Le componenti principali pesate geograficamente per la definizione di indicatori compositi locali, Rivista di economia e statistica del territorio. 2016. 33, 52.
33. Fusco E, Vidoli F, Sahoo BK. Spatial heterogeneity in composite indicator: A methodological proposal. Omega. 2017.

Chapter 4
Structure of a Net Positive Analysis for Supply Chain Social Impacts

Catherine Benoit Norris, Gregory A. Norris, Lina Azuero, and John Pflueger

Abstract Net Positive may well be the buzzword of this decade. Beyond the noise, it has the potential to be a transformational movement, helping businesses to redefine their role in society, their social purpose. As an idea, it simplicity and candor make it both extremely attractive and powerful. It poses a great question and sets a challenge: Can we give more to the environment and society than we take? To be Net Positive a company (and its supply chain) handprint needs to be greater than its footprint. The Net Positive Project and Harvard SHINE have worked to clarify the Principles and methodology that can make the Net Positive concept both actionable and valid. This include defining handprints in a measurable way. In this paper, we are developing on the methods that can be used to assess the social Net Positive impacts. Reviewing and building on social life cycle assessment, we introduce a structure for Net Positive analysis of social impacts. This framework is meant to be practical, actionable and inclusive.

4.1 Introduction

When the largest asset manager in the world (Blackrock) send a letter to CEOs of public companies asking them to start accounting for the societal impact of their businesses, you know this is a paradigm shift.[1] The Sustainable Development Goals perhaps paved the way for corporate actors to redefine their role in society, as contributors to improve social conditions. This in parallel to the UN business and human rights framework which established the need for companies to carry human rights due diligence and asked for greater accountability on their part.

---

[1] http://www.businessinsider.com/blackrock-ceo-larry-fink-just-sent-a-warning-to-ceos-every where-2018-1

C. B. Norris (✉) · G. A. Norris
Harvard School of Public Health, Cambridge, MA, USA
e-mail: catherinebenoit@fas.harvard.edu

L. Azuero · J. Pflueger
Dell Technologies, Round Rock, TX, USA

© The Author(s) 2020
M. Traverso et al., *Perspectives on Social LCA*, SpringerBriefs in Environmental Science, https://doi.org/10.1007/978-3-030-01508-4_4

There are several reasons for businesses to measure their supply chain and operation social impacts. For one, businesses need to assess their social and human rights risks and take steps to manage them. As corporate citizens, businesses have the opportunity to bring changes to improve social conditions and can be recognized by civil society for the positive changes they bring. Finally, companies may be formally appreciated for the (intrinsic) social value of their products. In this paper, we will focus on the first 2 identified needs.

The idea of Net Positive stems from the realization that if individuals and companies all have sustainability footprints (measurement of the cradle to grave negative impacts on the environment and society) they also have the capability to bring about changes that reduces those sustainability footprints and also creates what is called sustainability handprints (measurement of the cradle to grave positive impacts on the environment and society).

Net positive is defined as "putting back more to society and the environment than we take out." In short, it means: giving more than we take, or doing more good than harm.

Therefore, a Net Positive assessment entails the measurement of a product or organization footprint and also of its handprint.

For each impact category:

Net Positive = Handprint – Footprint

The Net Positive project considers handprint to be a change to business as usual. For organizations, business as usual is defined simply, as: responding to this year's demand with last year's products and processes.

The fact that handprints are all changes to business as usual means that companies are encouraged to create handprints in part by improving the life cycle performance (increasing the contributions and decreasing the footprints) of their products. Handprinting begins by optimizing the positive impact potential to be found within the company's core business.

The Net Positive project [1] has developed a set of principles for Net Positive that serves as a guide for companies seeking to integrate Net Positive in their strategy, goals or metrics development and that also provides a framework for Net Positive methodology development. The following figure presents the 4 Principles and their definition (Fig. 4.1).

4.2   The Scope

At the policy level, there is a consensus that a life cycle approach is needed to fully consider the environmental and social impacts of a product or an organization [2].

The United Nations Guiding Principles on Business and Human Rights (UNGP) [3] has also set the scope of social impact assessment to include anybody involved in or paid for goods and services in an organization 'operations or in its supply chain. This include value chain actors (suppliers), workers and their communities and

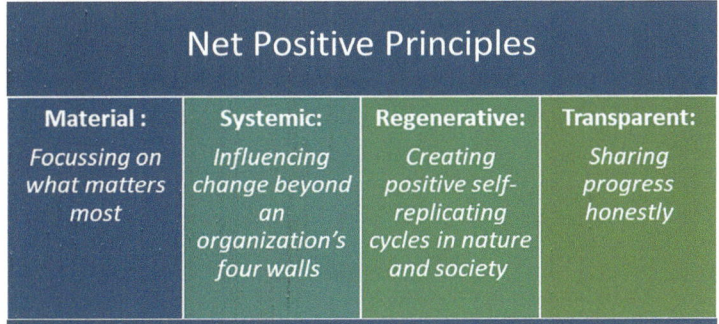

Fig. 4.1 Net positive principles

applies to all tiers, so also includes subcontractors, sub-suppliers and their workers and communities.

The Net Positive Project proposes a scaling process where organization increase the scope of their footprint and handprint assessment over time with the goal of:

- Implementing a full life cycle approach.
- Helping to select most beneficial production/ material and supply chain options
- Investing resources where they are most needed to address social hotspots and
- Creating the most beneficial scenarios for all stakeholders involved.

4.2.1 Social LCA

Social Life Cycle Assessment (S-LCA) is a technique used to assess the social and socio-economic impacts of products or organizations along their life cycle from extraction of raw materials to final disposal (cradle to grave approach).

S-LCA employs the modeling capabilities and systematic assessment process of LCA combined with relevant social sciences methods. The social aspects assessed in S-LCA are those that may affect stakeholders positively or negatively during the supply chain or life cycle of a product/ organization. The impact categories covered are largely defined by the international community through its policy framework and other social responsibility references and in respect to best available science (top down approach).

It can either be applied on its own or in combination with environmental LCA (*E*-LCA). It differs from other social impacts assessment techniques by its objects: products and services or organizations, and its scope: the entire life cycle. The scope (the life cycle) and the methodology (a systematic process of collecting and reporting about social impacts and benefits) are both key aspects that draw interest in the technique [4].

4.2.2 Positive Impacts in Social LCA

The question of positive impacts in Social LCA has been considered ever since its inception. The Guidelines for Social LCA express the aim of Social LCA as to improve social conditions in supply chains worldwide [5]. In particular, Social LCA is defined as a tool to support decision making [6].

This can include:

- Helping to select most beneficial production/material and supply chain options
- Investing resources where they are most needed to address social hotspots and
- Creating the most beneficial scenarios for all stakeholders involved.

However, in reality, very few journal articles have been published or guidance issued specifically on how to address positive impacts.

Our research using Harvard Hollis library system has returned only 3 journal articles focussing on the topic: one overall literature review on positive social impacts in Social LCA [7], one article discussing approaches and a proposal [8], and one focusing on societal value (societal benefits/ social utility, of a product during a product use phase) [9]. The two first articles review how positive impacts have been handled in case studies and in the literature.

Building on their analysis and adding ours, we argue that there are 5 main ways that the question of positive impacts has been handled so far:

- By identifying which subcategory of impacts has a positive connotation (eg. job creation) [8].
- By the way of performance reference points/ assessment scales where companies' performances are assessed from non-compliant to best practices. Scale levels after "compliance" are "positive" [10–13].
- By considering the absence of a negative impact to represent a positive impact (eg. no forced labour) [14, 15].
- By appreciating the social/ societal (intrinsic positive) value of the product (eg. vaccines, water treatment)
- By assessing where public health gains would be the highest by dollar purchased of an input/sector.

4.3 Our Framework

The assessment framework that we propose builds on the work of the net positive project and follows the same principles (material, systemic, regenerative and transparent).

Because supply chain is often where businesses' social hotspots are found, an approach reaching towards the full life cycle scope is needed. Therefore, our approach is based on social life cycle assessment. It positions change to business as usual as the key component of the approach. This is a new perspective for Social LCA because positive impacts have not been considered as change before.

Why change as the determinant of handprint creation? Because it is a clear indication of purposeful action to create positive impacts that goes beyond business as usual. Business as usual is determined based on the past year' activities. This recognizes businesses for taking an active role in the creation of social handprints.

Our approach is also integrating the UN Guiding Principles key recommendation of conducting human rights due diligence. The first step for businesses on their path to net positive is to calculate a social footprint and identify salient risks/ hotspots in order to assess where change is most needed (risks/ negative impacts) and establish a baseline that will support positive impact calculation.

Companies have several touchpoints by which they have leverage on social impacts. A company has tangible touch points with all main stakeholder categories and go beyond its own operations. For instance:

- The relationship between the business teams and suppliers.
- The relationship between the suppliers and their workers/ local communities/ other suppliers.
- The relationship between the ethical trade team and suppliers /workers.
- The relationship between the business teams/ lobbyists/ industry associations and governments.

In addition, a company can have leverage by:

- Joining other organizations in partnerships and collaborations;
- Using certified inputs if the certification manages effectively the social risks and create additional benefits;
- Designing products taking into account the risks and opportunities associated with the potential inputs.

As depicted in the following figure, companies' supply chains are complex and intricate. A business and each of its upstream suppliers has impacts on workers, communities, society, value chain actors and potentially, consumers (Fig. 4.2).

The first step for businesses on their path to net positive is to conduct a comprehensive materiality assessment. By comprehensive materiality assessment we mean that it should include stakeholders' qualitative assessment, potential impacts on the bottom line and social footprint results that ideally would be derived from what LCA refers to as a normalization analysis. It should include the identification of material negative responsibilities as well as material positive opportunities. A social footprint provides information about the impact categories most at risk for the company or its products, while normalization analysis compares these impacts to those of a larger "reference system" such as all activities in a given region for a given year; normalization results highlight which are the impact categories for which the industry (and its supply chain) accounts for higher contributions relative to their contributions on other impact categories. Normalization analysis identifies impact categories for which the industry has higher relative leverage than other impact categories. With

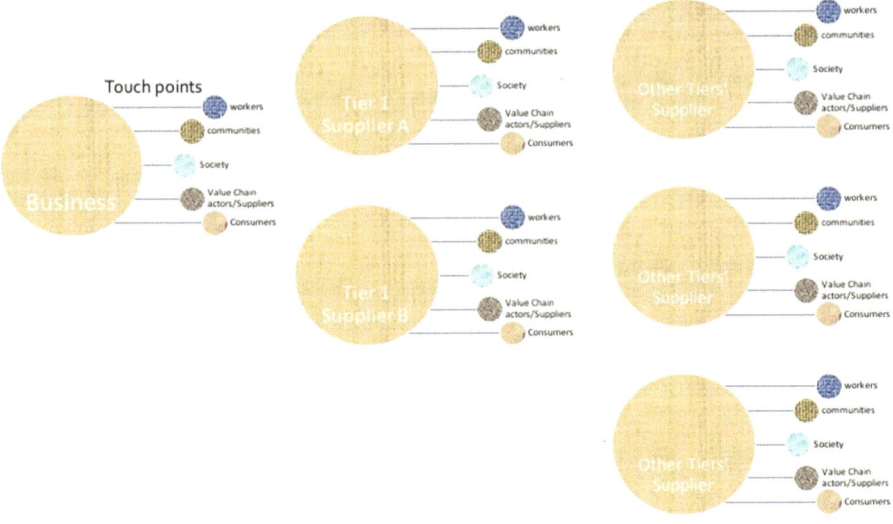

Fig. 4.2 Excerpt of a business universe of social impacts

the combined information provided by a full materiality assessment, it is possible to prioritize the impact categories that should be considered in the handprint assessment. For each category, a social hotspots assessment is conducted. This will identify which production activities and locations contribute a greater share to the total risk or impact for that category. The social hotspot assessment also provides the baseline against which progress will be measured. The next step is to collect site-specific information for the hotspot activities, to develop a refined baseline. This site-specific information is obtained by using social audit information, workers' survey results, participatory evaluation or other similar methods (Fig. 4.3).

Once the impact categories and the sites have been identified, a root cause analysis is required. A root cause analysis may be a simple exercise consisting of interviews or a more substantial research based on a literature review, stakeholder survey/interview and field research. A root cause analysis is necessary to evaluate what actions or changes will create the positive impacts desired. Sometimes the root causes may be related to business practices but sometimes the root causes may be related to cultural factors or conditions beyond those generally considered under the influence of a brand. Root causes can then be mapped to a company' touch points.

The creation of a handprint is the creation of a change, by implementing an intervention found to have leverage over the improvement of social conditions for an impact category. To measure the social handprint, we need to measure the outcome of the activity or *change* and its impacts related to the impact category (Fig. 4.4).

As an example, let's assume that child labor is a material impact for a business, and the business wants to reduce child labor risk at a supply chain location where there is currently a high risk. In order to reach a lasting solution to the problem there, we need to understand what are the drivers or cause of child labor. Is it poverty? Lack

Category
Baseline
Root cause
Change
Outcome
Impacts

Fig. 4.3 Handprint assessment

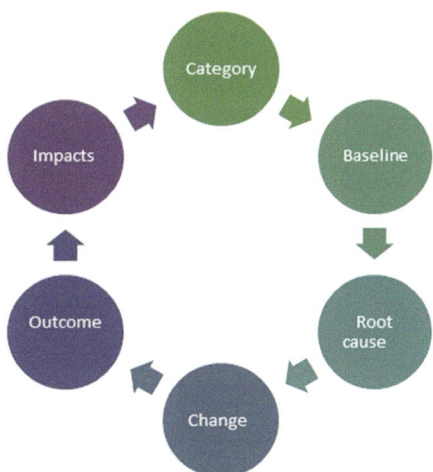

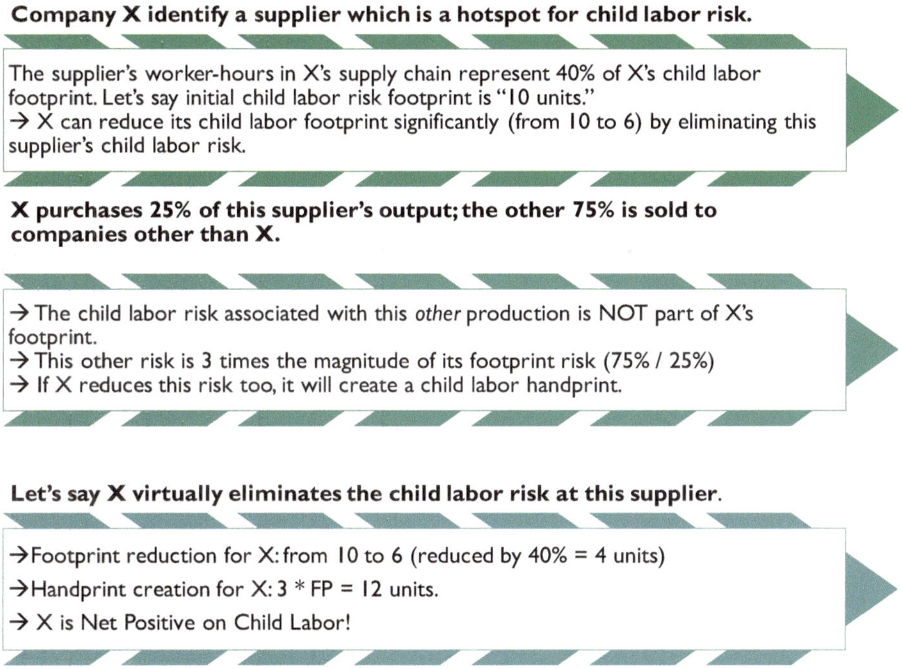

Company X identify a supplier which is a hotspot for child labor risk.
The supplier's worker-hours in X's supply chain represent 40% of X's child labor footprint. Let's say initial child labor risk footprint is "10 units."
→ X can reduce its child labor footprint significantly (from 10 to 6) by eliminating this supplier's child labor risk.
X purchases 25% of this supplier's output; the other 75% is sold to companies other than X.
→ The child labor risk associated with this other production is NOT part of X's footprint.
→ This other risk is 3 times the magnitude of its footprint risk (75% / 25%)
→ If X reduces this risk too, it will create a child labor handprint.
Let's say X virtually eliminates the child labor risk at this supplier.
→Footprint reduction for X: from 10 to 6 (reduced by 40% = 4 units)
→Handprint creation for X: 3 * FP = 12 units.
→ X is Net Positive on Child Labor!

Fig. 4.4 Example handprint calculation

of schools? Lack of a safe place to keep children? Lack of caretakers? Attempts to avoid child marriage? Strenuous daily production quotas?

When there is an understanding of what the root causes are, the action or "change" can then be planned. This may be a change in purchasing practices, or help/funding provided to suppliers to open a day care. After implementation we can measure direct outcomes, such as the number of children attending the day care or the reduction in daily quota strain. Finally, we can measure the impacts. These may be a reduction in child labor risk which can be expressed using the same units of measure used for social risk footprints, such as medium risk-hour equivalents, or similar. For the same impact category, footprint and handprint measures need to be calculated using the same unit. The handprint will account for the estimated total reduction in child labor risk, whether or not the reductions come within the business' footprint or not, and whether or not the activities affected participate within the business' supply chain or not.

4.4   Conclusions and Future Developments

Our framework is bringing several additions to what has previously been suggested by Social LCA scholars. Our research suggests that change is what a company can implement to generate positive impact. The question thus moves from identifying which impacts are positive or negative to what can be done to improve impacts on all categories. It also presents that a baseline calculation is always needed to serve as a basis to the impact calculation. Our framework also propose that we integrate the practice of root cause analysis to help companies to identify the actions that will bring lasting change. Measurement of social impacts, positive and negative is daunting. Many initiatives (Shared value, Social Capital Protocol) have been progressing developing concepts without fully working out their measurement. It is our hope that these proposals will bring the practice of social impact measurement a step further and that a community of practitioners will contribute to its future developments.

References

1. Norris GA. Net positive methodology. White Paper; 2017
2. UN Environment, Guidelines for Providing Product Sustainability Information. 2017. http://www.scpclearinghouse.org/resource/guidelines-providing-product-sustainability-information. Accessed 23 Jan 2018
3. United Nations. UN guiding principles on business and human rights. 2011. www.ohchr.org%2FDocuments%2FPublications%2FGuidingPrinciplesBusinessHR_EN.pdf&usg=AOvVaw1eXHpXS2jxinTbBidRBbsn. Accessed 23 Jan 2018
4. Benoît C, Norris GA, Valdivia S, Ciroth A, Moberg A, Bos U, Prakash S. The guidelines for social life cycle assessment of products: just in time! Int J Life Cycle Assess. 2010;15(2):156–63.
5. United Nations Environment Programme and Society for Environmental Toxicology and Chemistry. Guidelines for social life cycle assessment of products, Paris; 2009.
6. Jørgensen A. Social LCA, a way ahead? Int J Life Cycle Assess. 2013;8(2):296–9.

7. Di Cesare S, Silveri F, Sala S, Petti L. Positive impacts in social life cycle assessment: state of the art and the way forward. Int J Life Cycle Assess. 2018;23(3):406–21.
8. Ekener E, Hansson J, Gustavsson M. Addressing positive impacts in social LCA — discussing current and new approaches exemplified by the case of vehicle fuels. Int J Life Cycle Assess. 2010;23(3):556–68.
9. Shin KLF, Colwill JA, Young RIM. Expanding the scope of LCA to include 'Societal Value': a framework and methodology for assessing positive product impacts. Procedia CIRP. 2015;29:366–71.
10. Fontes J, Bolhuis A, Bogaers K, Saling P, van Gelder R, Traverso M, Tarne P, Das Gupta J, Morris D, Woodyard D, Bell L, van der Merwe R, Kimm N, Santamaria C, Laubscher M, Jacobs M, Challis D, Alvarado C, Duclaux C, Slaoui Y, Culley H, Zinck S, Stermann R, Carteron E, Gupta A, Nilsson S, Gaasbeek A, Goedkoop M, Evitts S,et al. Handbook for product social impact assessment version 3.0. 2016. http://product-social-impact-assessment. com/wp-content/uploads/2014/08/Handbook-for-Product-Social-Impact-Assessment.pdf., Accessed 23 Jan 2018
11. Revéret JP, Couture JM, Parent J. Socioeconomic LCA of milk production in Canada. In: Muthu S, editor. Social life cycle assessment, environmental footprints and eco-design of Products and Processes. Singapore: Springer; 2015.
12. Ciroth A, Franze J. LCA of an ecolabeled notebook – Consideration of social and environmental impacts along the entire life cycle. http://www.greendelta.com/uploads/media/LCA_laptop_ final.pdf., Accessed 23 Jan 2018
13. Ugaya CML, Brones F, Corrêa S, S-LCA: preliminary results of natura's cocoa soap bar, http:// www.lcm2011.org/papers.html?file=tl_files/pdf/paper/13_Session_Life_Cycle_Sustainabil ity_Assessement_I/6_Ugaya-S-LCA_Preliminary_results_of_Naturas_cocoa_soap_bar-763_b. pdf Accessed 23 Jan 2018
14. Traverso M, Finkbeiner M, Jørgensen A, Schneider L. Life cycle sustainability dashboard. J Ind Ecol. 2012a;16(5):680–8.
15. Traverso M, Asdrubali F, Francia A, Finkbeiner M. Towards life cycle sustainability assessment: an implementation to photovoltaic modules. Int J Life Cycle Assess. 2012b;17(8):1068– 79.

Chapter 5
Weighting and Scoring in Social Life Cycle Assessment

Breno Barros Telles do Carmo, Sara Russo Garrido, Gabriella Arcese, and Maria Claudia Lucchetti

Abstract Social impact evaluation is one of the cornerstones of products and services sustainability. Social Life Cycle Assessment (S-LCA hereafter) focuses on studying potential social impacts of products' life cycle. As it is a relatively new analytical approach, no globally shared application tools have been developed for it yet. Communicating S-LCA results to decision-makers in order to promote social sustainable decisions is a challenge because it involves the aggregation of companies' performances across impact categories through numerical variables based on value-choices. Currently, the weighting process (used for performance aggregation) considered for type I analysis in the literature presents some limits: lack of transparency, implicit choices, no standard weighting method and the failure to take into account the uncertainty of these value choices. This paper aims to address these limits by proposing a standard approach to conduct the weighting process for type I S-LCA. It starts after characterization phase and comprises four stages: (i) impact level scoring, (ii) functional unit aggregation, (iii) weighting factors definition and (iv) performances aggregation across impact categories. This approach is able to consider determinist or stochastic numerical variables, depending on the inclusion or not of the uncertainty associated to people' value judgments. In terms of results, this paper presents an illustrative case study in order to exemplify how to conduct the weighting process in S-LCA. Considering the results, we identified some limits related to our approach: (i) depending on the subjects involved in S-LCA and the subcategory indicators considered for the assessment, it might not be possible to define standard weighting factors for all case studies; (ii) the type of uncertainty tackled on this approach is only associated with value choices – no other

B. B. T. do Carmo (✉)
Universidade Federal Rural do Semi Arido – Engineering Centre, Mossoró, Brazil
e-mail: brenobarros@ufersa.edu.br

S. R. Garrido
CIRAIG, Montreal, Canada

G. Arcese
Università degli Studi di Bari Aldo Moro, Bari, Italy

M. C. Lucchetti
Roma Tre University, Rome, Italy

M. Traverso et al., *Perspectives on Social LCA*, SpringerBriefs in Environmental Science, https://doi.org/10.1007/978-3-030-01508-4_5

45

source of uncertainty is addressed and; (iii) the method used to assess qualitative social performances (scoring, check list or social hotspot database) can influence the aggregated social performance of product systems.

5.1 Introduction

Social Life Cycle Assessment (S-LCA) approach is useful to increase knowledge, clarify choices and promote the improvement of products' life cycle social conditions [1]. It can be also used for comparing among product systems, life-cycle stages or impact categories through an aggregation procedure – conversion and the possibly aggregation of indicator results across impact categories using numerical factors based on value-choices [2]. In essence, aggregation in S-LCA allows passing from inventory indicators results to a subcategory result and/or passing from subcategories results to an impact category result through the use of weights. For example, impact subcategories/categories that are deemed more important will have greater values attributed to them, to ensure that their associated results have a heavier weight on the final results than other results. As such, weighting can take place at any point in the study when aggregation takes place. The weighting process usually occurs while an aggregation of results is conducted. Aggregation allows to bring together separate results (e.g., subcategory indicator results) in order to boost interpretability – the aggregation can be performed across product systems, life-cycle steps, impact categories or stakeholder categories. Ekener-Petersen and Moberg [3] argue, however, that aggregation implies a loss of detailed information, highlighting the uncertainty associated with the definition of the weights [3].

Considering the nature of social assessments – (i) multidimensional indicators (each indicator is expressed in different qualitative/quantitative units), (ii) conflicting objectives (it is impossible to maximize performances for all indicators) and (iii) uncertainty associated with the performance assessment – practitioners should always keep in mind that disaggregated results should be presented along with the aggregated results, in order to avoid information being lost. Social phenomenon is multifactor and it can require the use of many dimensions to be measured. As such, it could happen that these dimensions cannot be aggregated and it should be preserved the importance of single dimensions in the weighting process.

After the social assessment, scientific literature currently conducts a weighting process in order to provide some aggregation and make easy the usability of this information into decision-making problems. In order to realize the aggregation process, it is necessary to establish S-LCA numerical parameters: weighting and scoring factors. Two types of variables are necessary in order to provide an aggregated social performance through life cycle approach [2]:

- Weighting factor: defined here as the accorded importance to the subcategory indicators pertaining to each stakeholder dimension or accorded importance for each stakeholder dimension when proceeding a complete aggregation;

- Scoring factor: defined here as a numerical score attributed to the classification levels (A, B, C and D) of each subcategory indicator.

For the first element (weighting factors), there is no a standard process in S-LCA domain to define them, as remarked by Russo Garrido et al. in 2016 [4]. They identified four different approaches used to define weighting factors: (i) equal weighting for subcategory, categories, unit processes or life cycle phases; (ii) worse performance prioritization; (iii) value judgments of stakeholders/experts/ users and (iv) norms and general literature. As such, different weighting factors sets may conduct to different rankings when comparing product-systems based on their "social performance". Considering these methods, we identified some limits of current approaches: (i) lack of transparency; (ii) implicit choices when defining weighting factors and (iii) no universally weighting method for representing the value judgments and none approach addressing the uncertainty associated to them.

Considering the second element (scoring factors), currently research in S-LCA considers the assumption of linearity when defining the scoring factors associated to the classification levels of the subcategory indicators, as identified by Carmo et al. [2]. For example, the majority of S-LCA papers consider the cardinal scale 1, 2, 3, 4 to translate qualitative assessments of product systems in each subcategory indicator into a numerical social score. However, this assumption cannot be guaranteed because each subcategory indicator is unique, presenting different performance reference points and contexts, as SAM (Subcategory Assessment Method), proposed by Sanchez-Ramirez et al. in 2014 [5]. As such, it is not reasonable allocate the same scoring factors to all subcategory indicators. Thus, customized scoring factors may be established for each subcategory indicator. How to define these scoring factors is a challenge, but the value judgment of S-LCA experts is an interesting starting point, as Carmo et al. [2] argued when proposing their methodology.

Considering these two variables types, we identified some limits into aggregation process used in S-LCA: lack of transparency, implicit choices, no standard weighting method and the failure to take into account the uncertainty of these value choices. Considering that both weighting and scoring factors can be defined based on value judgments and this approach is suitable because it is able to transform implicitly choices into explicit information, this paper aims to develop a method in this direction. However, it is necessary to model the value judgments' subjectivity and their variety because the recommendations may be modified when taking into account different group of people, as demonstrated by Carmo et al. [2].

As such, this paper aims to propose a standard approach to conduct the weighting process for type I S-LCA that will be available at the revised S-LCA UNEP/SETAC Guidelines (2009) being developed by the working group Social Life Cycle Alliance (SLC Alliance) [6].

5.2 Methodology

The approach proposed in this paper comprises six phases: (i) characterization; (ii) impact level scoring; (iii) functional unit aggregation; (iv) weighting factors definition; (v) performances aggregation across impact categories and (iv) interpretation.

The first phase is the qualitative assessment. Actually, there is a guideline suggesting a list of subcategory indicators that can be considered when assessing a social performance [1]. This paper does not address how to proceed the choice among the subcategory indicators.

The first phase is also the moment to define the qualitative scales through the Performance Reference Points (PRPs). Russo Garrido et al. (2016) [5] identified different approaches used for this purpose. All of them can be applied at our framework, but the method used for assessing social performances influences the aggregated social performance of product systems because different scales conduct to different evaluations. The importance at this point is to use the same assessment method when comparing product systems [5].

At the second phase, impact levels scoring, we must translate qualitative scales into quantitative scales. It can be used a linear scoring (for example: 1, 2, 3 and 4 to represent A, B C and D), as proposed by Sanchez-Ramirez et al. [7] or a customized scoring method, as proposed by Carmo et al. [2] [2, 7]. If the customized scoring factors were established based on value judgments, it is important to address this source of uncertainty. Carmo et al. (2017) [2] proposed a method for this purpose. They represented this uncertainty by probability density functions (PDF) based on a set of experts involved in the assessment. It can be also considered stakeholders' representatives or decision-makers. The translation of qualitative classification levels into quantitative numerical scores were done through a customized value function established from S-LCA expert judgments for each subcategory indicator. The PDFs were generated from S-LCA expert answers.

The scores obtained through the translation of qualitative assessment into quantitative social scores can be aggregated for all product life cycle (third phase of our methodology). For this purpose, the activity variable can be used. This paper does not touch this aggregation factor. We consider the traditional methods used by other researchers.

The performances obtained in different subcategory indicators can be aggregated by stakeholder dimension, for example. As such, it is necessary to define the weighting factors [8]. In this phase, weighting factors definition, we consider weighting concept as an intrinsic idea of defining priorities. Russo Garrido et al. also identified the most used approaches to conduct weighting in S-LCA [5]. Carmo et al. proposed a method to address the uncertainty in this phase when considering people's value judgments [2].

The weighting factors were defined based on the value judgment of the S-LCA experts – they distribute 100 points between the subcategory indicators pertaining to a given stakeholder dimension depending on their importance. Through Monte Carlo simulation, we generated the sets of weighting factors associated with each subcategory indicator, considering their point of view.

To conduct the final aggregation, the fifth phase of our methodology, considering the performances obtained by each product system along the whole life cycle and the weighting factors (stochastic or deterministic), we multiply these variables in order to calculate the aggregated social score.

As such, this approach is able to consider determinist or stochastic numerical variables (scoring and weighting factors), depending on the inclusion or not of the uncertainty associated to stakeholders, experts or decision-makers' value judgments.

Finally, considering the interpretation phase, it is important to make clear the subjectivity choices: how PRP scales were defined; the type of scoring used for the translation; the weighting factors considered for the aggregation and if/how uncertainty was addressed. Fig. 5.1 presents the schematic framework of our method.

5.3 Results

Here we present the type of results that can be obtained at each step of our approach. The aggregated social score (deterministic or stochastic) allows comparing four product systems by a single score. Fig. 5.2 and Fig. 5.3 illustrate the type of results we obtained considering an illustrative case study. For more details, consult Carmo et al. [2] [2]. For stochastic results, we only present here the graph for the workers stakeholder dimension.

The results presented by Figs. 5.2 and 5.3 show that for the social score representing Workers stakeholder dimension, deterministic ranking (PS3 > PS1 > PS4 > PS2) is the same that the ranking provided by implementing stochastic data. However, we cannot guarantee that always stochastic results will provide the same answer that deterministic results because if the value judgments vary a lot, there is an overlapping of the PDFs representing the aggregated social score. It is important to remark that the overlapping can be much more important than the one we observed at Fig. 5.3, as demonstrated by Carmo et al. [2].

5.4 Conclusions

Different methodologies in scientific literature can be used to assign the importance to subcategory indicators when aggregating results in S-LCA. This paper aimed to propose a framework able to organize the aggregation for this purpose.

Although the reference point to evaluate some social aspects of products is represented by the S-LCA guidelines (UNEP/SETAC 2009), the S-LCA methodology follows ISO 14040-44 standards [9], which are available for the Environmental LCA. Further development for the social dimension is still needed.

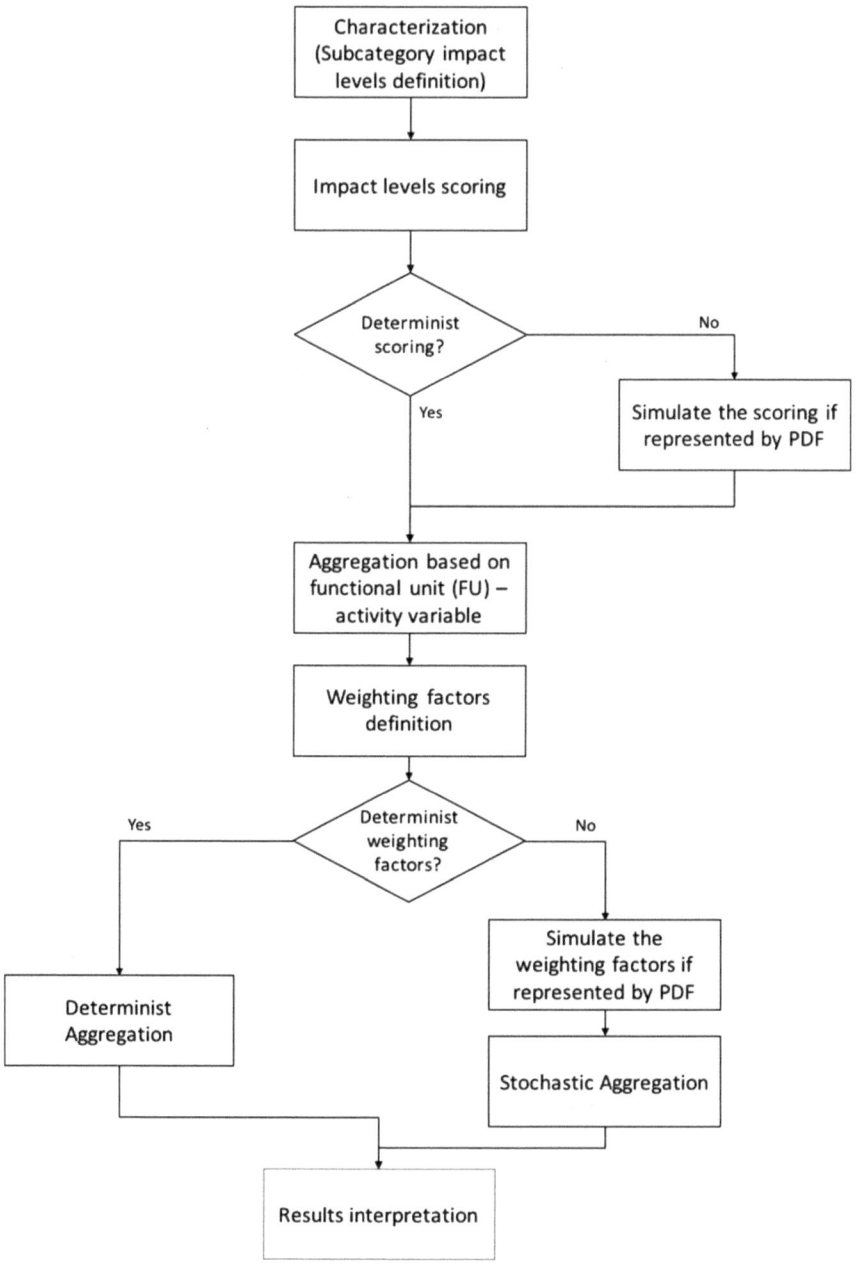

Fig. 5.1 Framework to conduct aggregation into Social Life Cycle Assessment

Fig. 5.2 Types of results obtained applying our approach – Deterministic results

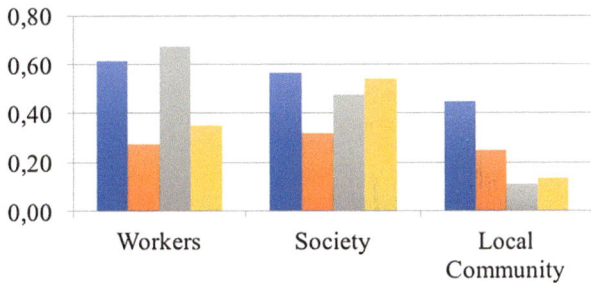

Fig. 5.3 Types of results obtained applying our approach – Stochastic results

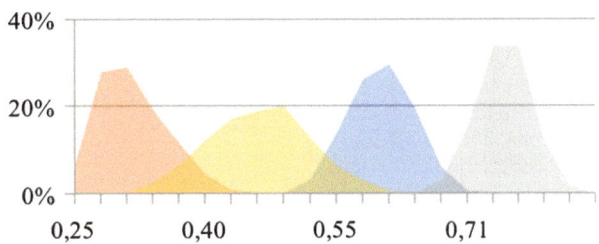

As such, in the framework presented, we argue the necessity to be transparent in aggregation process, proposing general steps to follow for the S-LCA Guideline revision. This paper addressed two key points for aggregation procedure: weighting and scoring factors.

Currently research in S-LCA defines weighting factors based on the value judgment of decision-makers or S-LCA experts. The point of view of the group is not the same for the majority of the cases and as such sensitivity analysis must be conducted.

On the other hand, currently research in S-LCA considers the assumption of linearity when defining the scoring factors associated to the classification levels of the subcategory indicators. However, this assumption cannot be guaranteed because each subcategory indicator is unique, presenting different performance reference points and contexts, as SAM (Subcategory Assessment Method), proposed by Sanchez-Ramirez et al. [7]. As such, customized scoring factors may be established for each subcategory indicator.

Our approach is able to provide aggregated social scores for each stakeholder dimension. We identified some hotspots applying it: (1) the weighting factors are customized for each case study because it depends on the subjects involved in S-LCA and subcategory indicators considered for the assessment; (2) the uncertainty is able to represent only the value choices variety – no other source of uncertainty was addressed and; (3) the method used for assessing qualitative social performances (scoring, check list or social hotspot database) influences the aggregated social performance of product systems.

Finally, when presenting the results, it is important to make clear the subjective choices: how PRP scales were defined; the type of scoring used for the translation of qualitative assessments into scores; the weighting factors considered at aggregation and if/how uncertainty was addressed.

References

1. Benoit C, Norris GA, Valdivia S, Ciroth A, Moberg A, Bos U, Praksha S, Ugaya C, Beck T. The guidelines for social lifecycle assessment of products: Just in time! Int J Life Cycle Assess. 2010;15(2):156–63.
2. Carmo BBT, Margni M, Baptiste P. Addressing uncertain scoring and weighting factors in social life cycle assessment. Int J Life Cycle Assess. 2017;22(10):1609–17.
3. Ekener-Petersen E, Finnveden G. Potential hotspot identified by social LCA – Part 1: A case study of a laptop computer. Int J Life Cycle Assess. 2013;18(1):127–43.
4. Ugaya CML. The social assessment of products. In: Murray J, Mcbain D, Wiedmann T, editors. The sustainability practitioner's guide to social analysis and assessment. Chicago: Commun Group; 2015. p. 18–27.
5. Russo Garrido S, Parent J, Beaulieu L, Revéret JP. A literature review of type I S-LCA – making the logic underlying methodological choices explicit. Int J Life Cycle Assess. 2018;23(3):432–44.
6. https://www.social-lca.org/. Accessed DD.MM.2018
7. Sanchez-Ramirez PK, Petti L, Haberland NT, Ugaya CML. Subcategory assessment method for social life cycle assessment. Part 1: methodological framework. Int J Life Cycle Assess. 2014;19 (8):1515–23.
8. Bengtsson M. Weighting in practice: Implications for the use of life-cycle assessment in decision making. J Ind Ecol. 4(4):47–60.
9. ISO 14040. Environmental management – life cycle assessment – principles and framework. Geneva, 2006.

Chapter 6
Beyond a Corporate Social Responsibility Context Towards Methodological Pluralism in Social Life Cycle Assessment: Exploring Alternative Social Theoretical Perspectives

Henrikke Baumann and Rickard Arvidsson

Abstract The UNEP/SETAC guidelines have Corporate Social Responsibility (CSR) as the underpinning theoretical perspective. However, studies on CSR suggest that the companies have benefitted more than society. We explore two alternative theoretical perspectives: the theory of ecologically unequal exchange (TEUE) and the actor-network-theory (ANT). By analysing case studies informed by TEUE and ANT, we identify their contribution to social life cycle assessment. The analysis shows that the perspectives enable description and identification of issues otherwise uncovered by the UNEP/SETAC approach: the unequal balance of health effects over a production and a consumption system and the presence of multiple and sometimes conflicting interests across actors in a production and consumption system, respectively. We point out characteristic methodological differences and conclude that S-LCA would benefit from greater pluralism.

6.1 Introduction

Most current efforts in Social Life Cycle Assessment (S-LCA), and in particular the UNEP/SETAC guidelines, have Corporate Social Responsibility (CSR) as the underpinning theoretical perspective. The basis in CSR is reflected in, for example, the similarity of key terminology (e.g. stakeholder as in stakeholder categories). CSR is a corporate self-regulatory mechanism that urges companies to go beyond self-interest and beyond legal requirements regarding ethical, environmental and social standards. Although CSR is a useful and legitimate response to the sustainability challenges facing companies, there are also limitations to what CSR can achieve. Over 50 years of studies on CSR suggest that the companies themselves have benefitted more than has society at large [1, 2]. In short, CSR has been criticised for legitimising and consolidating the power of large corporations and for advancing

H. Baumann (✉) · R. Arvidsson
Environmental Systems Analysis, Chalmers University of Technology, Göteborg, Sweden
e-mail: henrikke.baumann@chalmers.se

© The Author(s) 2020
M. Traverso et al., *Perspectives on Social LCA*, SpringerBriefs in Environmental Science, https://doi.org/10.1007/978-3-030-01508-4_6

sustainability efforts that benefit the companies and not necessarily sustainability at large.

Taking a step back, one may contemplate what does the word 'social' mean? What does it mean when some activities have a 'social dimension'? There is no one simple answer to these questions. According to the Oxford English Dictionary, social science is the scientific study of human society and social relationships. It covers a broad range of fields, not only sociology but also political science, economics, social and cultural anthropology, law, among others.

The social sciences differ greatly from the natural sciences in that there are few, if any, genuine law-like causal regularities that govern social phenomena [3]. This lack is sometimes explained by the complexity of human behaviour and the social world, and there is presently no agreement about the proper approach to investigating the social world. This is reflected by the methodological pluralism in social inquiry [3, 4]. Kauffmann [5] concludes that "methodological pluralism is a necessary characteristic of sustainability science as a whole". Here, pluralism is within the scientific enterprise itself, and should not be confused with social pluralism such as stakeholder pluralism in CSR [4]. In line with this, we argue that a social perspective on products does not have to be limited to a social description determined by a corporate point-of-view and that the many facets of the 'social' require greater methodological pluralism in the field of S-LCA. Here, we explore two different approaches, both theoretically-informed from different fields within the social sciences. In addition, we provide some contrasting observations on these studies had conventional S-LCA (i.e. following the UNEP/SETAC guidelines [6]) been used.

6.2 Variations on Social Product Studies

In our research, our focus has been on conducting socially relevant life cycle studies. In doing so, we have had to depart from the UNEP/SETAC guideline for S-LCA [6]. Two theoretical bodies have been particularly useful for informing our S-LCAs: the theory of ecological unequal exchange (TEUE) and actor-network-theory (ANT).

The theory of ecological unequal exchange (TEUE) describes the unequal material exchange relations and consequent ecological interdependencies within the world economy, all of which are fundamentally tied to wide disparities in socio-economic development and power embedded within the global system [7, 8]. It has mainly been applied to the ecological analysis of trade between countries, but some analyses at product level using LCA can also be found [cf. 9]. Here, we draw on TEUE to look at the balance of social impacts in terms of disability-adjusted life years (DALY) for different parts of the production and the consumption system. The focus on unequal exchange in TEUE means that an analysis looks into the balance between two sides. While Oulu [9]analysed the (im)balance of environmental impacts related to two products per the volume of trade between two countries, we looked into the (im)balance of social impacts between the negative impacts found in

Table 6.1 Overview of conducted social studies of product systems informed by TEUE and ANT

Theory of ecologically unequal exchange	Actor-Network-Theory
Airbag system, by Baumann et al. 2013 [14]	Cocoa supplies, by Afrane et al. 2013 [16]
Wedding ring, by Arvidsson et al. 2018 [15]	Shrimp production and consumption in Sweden, by Camacho Otero & Baumann 2016 [17]
Catalytic converters, by Arvidsson et al. 2018 [15]	Metal packaging and extended producer responsibility in Sweden and the Netherlands, by Lindkvist& Baumann 2017 [18]

the production system and the positive impacts related to the product function in the consumption system.

Actor-network-theory (ANT) is an approach for exploring how networks are built or assembled and maintained to achieve a specific objective [10]. Borrowing the words of Czarniawska[11], ANT "is not so much a theory of the social as a suggestion for how to study the social". One of the characteristics for ANT is that 'actors' denote both human and non-human actants. ANT was originally developed within the social studies of science and technology, and it has come to be applied in many fields and disciplines. Product Chain Organization (PCO) is a life-cycle-related application of ANT used for the study of the actor-networks shaping product flows [12, 13].

We have applied TEUE and ANT to several product studies (Table 6.1). In the following, we illustrate their respective contribution to social product studies by describing their application to four of these studies in greater detail.

6.2.1 Airbag

In this product study [14], the company at hand was interested in a S-LCA addressing the social rationale for one of their key products. The product was an airbag system. Given that the purpose of an airbag system is to prevent injuries and fatalities, we sought relevant ways for describing such impacts. Most cars are now fitted with a number of airbags. The number differs for each car model and maker. Some cars have up to 23 airbags to protect the driver and passengers in frontal and side collisions; many have fewer. An airbag system consists of a number of sensors in a vehicle that send information to an electronic control unit (ECU), which in the event of a collision triggers various firing circuits to deploy one or more airbag modules. Such airbag modules are deployed through a pyrotechnic process.

In our study, we described one airbag module [14]. Eventually, we came to use DALY as indicator since it covered both injuries and lives lost during production as well as injuries and lives saved during use. We found that the largest DALY losses stemmed from electricity production, followed by toxic emissions in mainly electronics production. The production of explosives and mining of metals contributed with the lowest DALY loss in the product system. Since the DALY saved were about 300 times higher than the DALY lost for a single airbag module, the results indicate that the purpose of an airbag system, which is to save lives and prevent injuries, may be socially justified.

Drawing on TEUE, we also analysed the distribution of DALY along the product life cycle. This analysis pointed to an unequal distribution of the benefits and the harms of the airbag systems. The greatest benefits are to be found for users of expensive cars with many airbags, mainly in the Global North, whereas the majority of the harms are to be found with workers in electronic and energy (coal) production in mainly the Global South.

Had conventional S-LCA methodology been applied, it is likely that such a study would have explored only the negative impacts in the production of an airbag system. Thereby, the socially unequal distribution of benefits and harms related to the airbag system would have been difficult to identify.

6.2.2   Catalytic Converters

Looking further into car-related safety measures, the next product to investigate by the TEUE-inspired DALY approach became a catalytic converter [15]. Their aim is to convert the toxic air pollutants carbon monoxide (CO), nitrogen oxides (NOx) and hydrocarbons into harmless substances. The converter consists of a ceramic honeycomb substrate to support the catalyst, an insulating material, a heat shield, a steel housing, wash coats and the platinum group metal (PGM) catalyst itself. Different scenarios were explored regarding emission conversion efficiency for CO and hydrocarbons, functional lifetime (100–200, 000 km), PGM recycling rate (3–50%) and geographical locations. Although the PGMs only accounted for 0.006 kg of the total weight of 7.3 kg, they accounted for most of the health impact. Whereas occupational accidents had a minor contribution, the assessment of emissions caused during PGM mining versus emissions saved during use determined the net health impact of the converter. Given a time horizon of 20–100 years, DALY saved during use dominated. Given an infinite time horizon, emissions from PGM mining dominated for most scenarios, often resulting in a net negative health impact.

The study also highlighted the unequal distribution of impacts over the converter's life cycle. Benefits mainly occurred in the Global North where the converters are used, whereas impacts mainly occurred in proximity to the PGM mining, i.e. in South Africa or Russia. Again, this unequal distribution would not have been detected by the conventional UNEP/SETAC approach to S-LCA, where the emission reductions would typically not have been included.

6.2.3   Cocoa

The context for our study [16] was that in 2010, the company Unilever committed to source all its cocoa for a certain product range from sustainable sources within 5 years. Since companies like Unilever procures its chocolate from wholesalers, it has no direct contact with farmers. Certification thus provides a means of assurance that farmers adhere to a number of good agricultural practices. For this, Unilever partnered with Rainforest Alliance.

In our study, we described the network of actors in cocoa product chains in order to explore the effect of introducing new actors related to certification to a conventional product chain [16]. For practical reasons, the study was geographically limited to cocoa grown in Ghana, the world's second largest producer of cocoa. Product chain actors were identified through multiple sources and on-site in Ghana. The interviews covered each actor's role and relationships, which enabled the mapping of the PCOs. Additionally, qualitative interviews explored actors' views and perspectives on sustainability and certification in order to understand for premises for sustainability in the chain. Visits to three farming regions provided a rich and diverse sampling of viewpoints and farming practices.

We found that the cocoa industry involved diverse actors (see Fig. 6.1), linking a multinational corporation to numerous smallholder farmers, who typically operate on approximately 1 ha of land, are often poor and illiterate, and production is often a family effort. In Ghana, there are also governmental bodies regulating the national cocoa industry. In this case, certification with Rainforest Alliance came with training

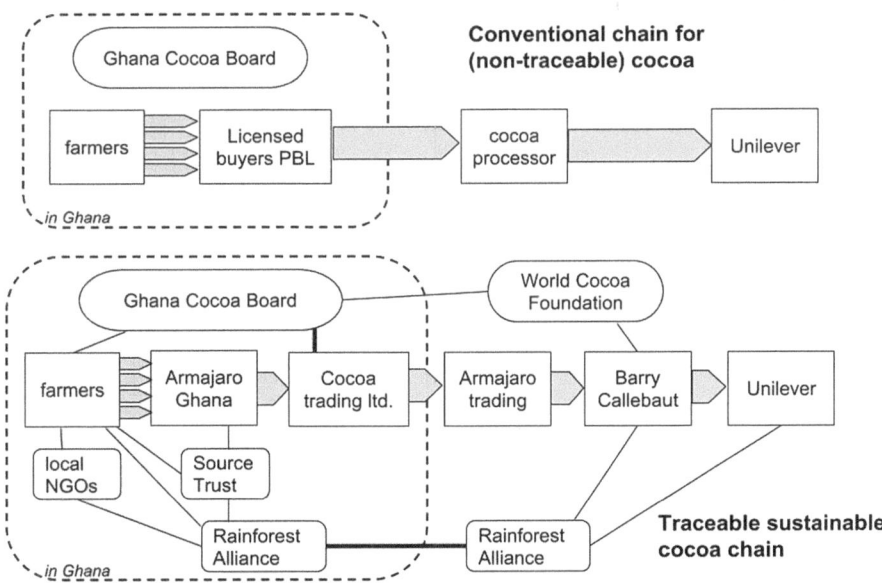

Fig. 6.1 Overview of the product chain organisation for conventional and certified cocoa chains

Table 6.2 Overview of viewpoints communicated by actors in the cocoa product chain, where some are agreed upon, others are conflicting, and some are held by a single actor only

Strong agreement among chain actors	Views expressed by only certain actors ('lone voices')	Conflicting views among chain actors
Deforestation	Climate change (farmers only)	Child labour concerns (some farmers and other actors believe child labour and trafficking has been addressed and is being eliminated through certification, whereas others believe it is still a problem)
Soil depletion	Landowner system and conflict of land (Ghanaian governmental bodies only)	
Farmer income		
Productivity		
Community development	Illegal logging, illegal mining, slash and burn (Ghanaian actors only)	
Lack of education and knowledge	Critique of certification schemes and preference for collaborative efforts for socio-economic development (Ghanaian governmental bodies only)	Food safety issues (some actors see this as an important concern while other believe it has been tackled well)

for the farmers. Such training is not always the case in certification but was made part of it to secure more sources to sustainable cocoa for Unilever.

The PCO showed that sustainability views are not uniform throughout the product chain (see Table 6.2). While there is strong agreement on certain issues, views differ substantially on others or are limited to few actors. For most cocoa chain actors, environmental issues were secondary. The exception were the farmers who realised the reality of climate change and its adverse impact on cocoa farming, and Rainforest Alliance which also had strong views on deforestation and biodiversity. Many had positive views on certification, but negative concerns were also expressed. Some actors were confused about the many, competing certification schemes (e.g. Rainforest Alliance, UTZ and Fairtrade) and showed resistance to the entailing administrative work of handling multiple certification schemes. More specifically, the governing cocoa bodies in Ghana suggested that sustainability could be improved without certification and would have preferred that sustainability efforts were organised more collaboratively.

The switch from conventional to traceable, certified cocoa sources at a multinational corporation led to changes in the structure of the cocoa industry in Ghana. The analysis brings to light the multiple and sometimes conflicting views on the development of the product chain towards greater sustainability. Perhaps most significant among these are the concerns expressed by Ghanaian governmental bodies for the socio-economic development of the cocoa industry and the limits to their self-determination related to the format given by certification schemes advanced by multinational corporations.

Had conventional S-LCA methodology been applied, it is likely that semi-quantitative subcategory indicators would have been used without an in-depth understanding of the production system. Such a study would likely also have focused on stakeholders to the certification initiative at hand, thereby producing a positive but limited picture of its effects. A typical corporate perspective would thus have neglected the multiplicity of actor viewpoints on sustainability and the tensions around certification schemes for sustainable socio-economic development for small-holder farmers.

6.2.4   Shrimp Production and Consumption in West Sweden

The study about the Swedish West coast shrimp was prompted by a controversy about its eco-labelling and sustainability assessment. Controversy mapping was carried out as a way of looking into the matters of concern at hand. The mapping was subsequently analysed through a life cycle lens [17].

Controversy mapping is a methodology developed in the scientific humanities [19, 20]. It provides a possibility to look into matters of concern as key realms for social construction. In contrast to matters of fact, matters of concern are unfinished issues under construction by many actors that interact through various devices [10]. The method involves the tracing of statements, literatures, and actors drawn in into a controversy. By assembling these elements, it is possible to describe the process of the controversy as it evolves and to identify matters of concern over which networked actors 'wrestle'. Using softwares for network visualisation and analysis, the extent of actor-networks can be shown and the roles of actors be analysed.

The controversy started in 2014 on the Swedish West coast when the Swedish WWF issued a red light for the locally fished shrimp in their seafood consumer guide. The controversy played out in the public realm, in the news and the social media. People put forward viewpoints, contested those of others, referred to documents, enrolled and mobilised others, took action, etc. The red light issued by WWF found subsequent support when the Coast Guard apprehended a fishing vessel illegally dumping shrimp in the middle of the sea and when the Swedish branch of IUCN red-listed the shrimp, but it was also countered by increases in fishing quotas and the issuing of a Marine Stewardship Council eco-label obtained by the local Fish Auction. A temporary stabilisation was reached in May 2016 when WWF endorsed the MSC eco-label while keeping the red light for the shrimp at the same time.

Looking at how the controversy played out over the life cycle, it is found that many of the product chain actors are present in the public debate, albeit to varying extents. What is also seen is a divide between actors in the consumption system and the production system. The action of red-lighting by WWF is directed at consumers, and it also received wide support from actors in the consumption system. However, this led to many reactions in the production system, where the opposing viewpoints became concentrated (see Fig. 6.2). This also led to the application for MSC certification by a central actor in the production system.

To begin, the controversy apparently revolves around the sustainability of shrimp fishing. Central was the interpretation(s) of the scientific data on shrimp population size, where the two opposing sides in the controversy referred to the same report and data. However, the controversy also gets connected to topics of culture, livelihoods and the traceability of products. Going deeper into the empirical material, it becomes

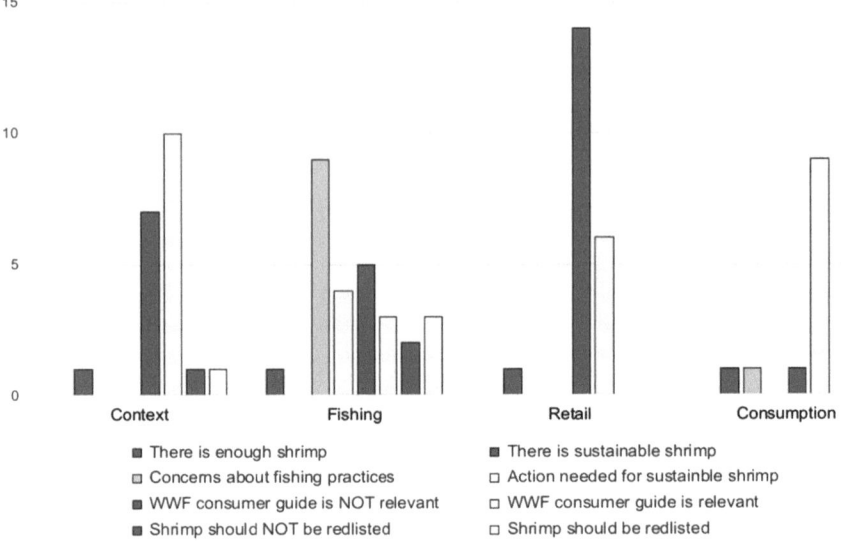

Fig. 6.2 Viewpoints across product chain positions. Light bars for viewpoints expressing concern about the sustainability of the shrimp; dark bars for opposing viewpoints

possible to see that a large part of the disagreement centres on the legitimacy of one actor's call to stop consumption of shrimp from a particular stock. This part of the dispute revolved around WWF's legitimacy and touches upon trust in business and governmental institutions in society. In addition, a smaller controversy is nested inside, one that centres on the fact whether or not there is enough shrimp in the sea. This part of the dispute involved the 'louder' voices from the government and the industry who were arguing about the accuracy of stock assessments related to red-lighting, red-listing and fishing quotas, which is a debate about science and the interpretation of evidence.

Had a conventional S-LCA following the UNEP/SETAC guidelines [6] been conducted, it is likely that it would have been conducted by an industry actor in the production system where statements about livelihoods, among else, were made. However, controversy mapping identified the problematic position of those actors in the unfolding events. In the context of the controversy, an industry commissioned study would likely have been considered at best limited and lacking, if not biased, since conventional S-LCA methodology has great difficulty registering diversity in viewpoints and matters of concern.

6.3   Analysis

To identify the contribution of TEUE and ANT to S-LCA, we characterize what is the objective of the studied analyses and characterize the enabling methodological features relative to the current UNEP/SETAC S-LCA framework [6].

Two studies have been described in which DALY was used for impact assessment within a TEUE framework, the S-LCAs of an airbag system [14] and a catalytic converter [15]. These studies aim at assessing each product's contribution to social sustainability in the light of the whole life cycle. The purpose of these products is to diminish harms to health and even save lives. Since these product qualities are critical to the companies producing and selling them, the studies were designed so that each product's contribution to social sustainability in the light of the whole life cycle could be evaluated. This means that the product in itself needed to be socially characterized. In addition to being the basis for the functional unit in the study, it's social impact is calculated and evaluated in context of the rest of the life cycle. Having a functional unit as an 'active' impact parameter in the life cycle calculations is a marked difference to what is described in the current UNEP/SETAC S-LCA framework [6].

The product's functions made health-oriented indicators appropriate, and the DALY indicator was imported from ELCA methodology, rather than using any suggested in the UNEP/SETAC S-LCA guidelines. The social contribution of each product was analysed via contrasting positive impacts (saving of lives) in the consumption system with the negative impacts (harm to life) in the production system. The DALY indicator was useful here since it can both capture the positive impacts of safety and abatement technologies and the negative impacts of pollutants and working conditions. Matters of life and death are socially significant wherever this happens, and the calculation of a DALY balance over the production and consumption system showed that a positive social contribution of a product is not necessarily self-evident. In addition, TEUE provided a framework through which north-south inequalities in the DALY balance were put in sharp focus, thereby enabling a sustainability assessment of the impacts. Such a distinction between description and sustainability assessment is not clear in the current UNEP/SETAC S-LCA framework.

Two studies have been described where ANT provided a basic and useful framework for S-LCAs, cocoa product chains with and without certification [16] and a shrimp production and consumption system [17]. The aim of both the cocoa and the shrimp studies was to better understand the processes organising and shaping the product flow with regard to its socio-ecological sustainability. The focusing on understanding these product chains required great emphasis on description, rather than on assessment. As already mentioned, such a distinction between description and assessment is not clear in the current UNEP/SETAC S-LCA framework [6]. Moreover, in both studies, the products are associated with environmental problems and the social study of their life cycles provided important insight towards their sustainability. This brings to the fore the value of social studies of product life cycles also for environmental sustainability.

Since people in every community have preferences and make trade-offs [21], it was appropriate to view and hear them as actors in their own right, rather than as stakeholders in a CSR context. Also, in order to understand the organising of the product flow, it was necessary to also include other relevant and contextual actors influencing to the product flow, such as governing bodies, NGOs and celebrities.

Such a social scope is wider than what is currently suggested in the UNEP/SETAC S-LCA framework, which concentrates on those in direct contact with the product flow.

The broad aim in these two studies required a methodology that was both flexible and open so that significant issues could be identified and captured. This meant that the approaches needed to distinguish between matters of concern and matters of fact. For this, the two studies used somewhat different approaches to data collection. In the cocoa study [16], a qualitative approach was chosen with open-ended questions and field observations, enabling actors to freely describe their role and interactions with other product chain actors, as well as express their viewpoints on their situation and the state of things. For the shrimp study [17], a great amount of data was readily available since the controversy played out in the public realm in news and social media. Here, controversy mapping methodology was used to capture the entirety of viewpoints, actors and networks—this material was subsequently used for the social analysis of the shrimp life cycle. Especially for the identification of the matters of concern in the studied communities, it would have been impossible to use predefined indicators of the type listed in the UNEP/SETAC framework for S-LCA.

6.4   Discussion

Considering sustainability and how that can be achieved is really to think about sustainability governance [22]. Here, CSR cannot be the only approach. Testing other approaches to S-LCA becomes thus important.

Since a major concern of the social sciences is with the adaptation, preferences and trade-offs in communities, it matters how social assessment exercises take these into consideration [21, 23]. In the field of Social Impact Assessment, the problem of 'death by a 1000 cuts' refers to the problem of many individual activities being for the most part insignificant, but many taken together and over time, they may have major repercussions on a community's resilience. This problem has significance for S-LCA methodology in several ways. One is the need for a contextualisation of described impacts in an individual study so that a sustainability assessment is possible. Another one concerns how and when the life cycle perspective offers value to social studies. A third deals the difference of considering people as actors in their own right or stakeholders in a CSR context.

Currently, CSR-based S-LCA lacks contextualisation of impacts to make a sustainability assessment. With ANT-based S-LCA, the identification of matters of concern is possible, thereby offering a way to gauge important matters for a community with regard to its resilience. With DALY and TEUE, S-LCA becomes ultimately concerned with life and death, which are assessment criteria that matters everywhere, in any community. Here, it becomes meaningful to assess the entire life cycle using cumulative indicators with a mindful eye for inequalities.

We have here shown and discussed how different theoretical frameworks inform S-LCA and sustainability assessment in different ways. Allowing for such methodological pluralism is befitting S-LCA, given the complexity of human behaviour and the social world (Gorton 2006), and given the theoretical and methodological richness of the social sciences.

6.5 Conclusions and Future Outlook

The analysis and discussion showed that S-LCA informed by TEUE and ANT add to the current UNEP/SETAC framework in that description of phenomena and issues hitherto uncovered by the UNEP/SETAC framework is made possible. Some differing methodological features have also been described. Whether people in the product system should be considered actors in their own right or stakeholders to a CSR initiative marks a telling difference in our approaches compared to the UNEP/SETAC approach to S-LCA. Moreover, our studies recognize a difference between emphasis on the description of social aspects (i.e., the inventory) and their assessment according to certain social sustainability criteria (i.e., the impact assessment). These differing emphases could be viewed as two types of S-LCA, which could be characterised as social life cycle analysis and life cycle social sustainability assessment, respectively.

Given the complexity of human behaviour and the social world [3], we find that the field of S-LCA would benefit from greater methodological pluralism. Richer descriptions with multiple social perspectives provide better 'roadmaps' for advancing sustainability than assessment made from a unilateral perspective. Such pluralism could perhaps also sensitize corporations and CSR professionals to the many and sometimes competing perspectives, and contribute to a more nuanced understanding of how the social 'fabric' shapes the sustainability of the product chain they are part of.

References

1. Banerjee SB. A critical perspective on corporate social responsibility: towards a global governance framework. Crit Perspect Int Bus. 2014;10(1–2):84–95.
2. Sanders P. Is CSR cognizant of the conflictuality of globalisation? A realist critique. Crit Perspect Int Bus. 2012;8(2):157–77.
3. Gorton WA. The philosophy of social science. The Internet Encyclopedia of Philosophy — A peer-reviewed academic resource http://www.iep.utm.edu/soc-sci/#H1, Accessed 17 Dec 2017.
4. Keating M, Della Porta D. 2009. In defence of pluralism. Combining approaches in the social sciences. Political studies association, Edinburgh. 2009.
5. Kauffman J. Advancing sustainability science: report on the international conference on sustainability science (ICSS) 2009. Sustain Sci. 2009;4(2):233–42.
6. UNEP-SETAC. Guidelines for social life cycle assessment of products. United Nations Environment Programme, 2000.

7. Hornborg A. Zero-sum world: challenges in conceptualizing environmental load displacement and ecologically unequal exchange in the world-system. Int J Comp Sociol. 2009;50(3–4):237–62.
8. Jorgenson AK. Environment, development, and ecologically unequal exchange. Sustainability. 2016;8(3):227.
9. Oulu M. The unequal exchange of Dutch cheese and Kenyan roses: introducing and testing an LCA-based methodology for estimating ecologically unequal exchange. Ecol Econ. 2015;119:372–83.
10. Latour B. Reassembling the social: an introduction to actor-network-theory. Oxford: Oxford university press; 2005.
11. Czarniawska B. Actor-Network Theory. In: The SAGE handbook of process organization studies. London: SAGE; 2016. p. 160–72.
12. Baumann, H.Simple material relations handled by complicated organisation or 'how many (organisations) does it take to change a lightbulb?', Proceedings of what is an organization? Materiality, Agency and Discourse, HEC Montréal, Université de Montréal, Queébec, Canada, 2008.
13. Baumann H. Using the life cycle approach for structuring organizational studies of product chains. Linköping: Greening of Industry Network conference; 2012.
14. Baumann H, Arvidsson R, Tong H, Wang Y. Does the production of an airbag injure more people than the airbag saves in traffic? J Ind Ecol. 2013;17(4., 2013):517–27.
15. Arvidsson R, Hildenbrand J, Baumann H, Islam KN, Parsmo R. A method for human health impact assessment in social LCA: lessons from three case studies. Int J Life Cycle Assess. 2018;23(3):690–9.
16. Afrane G, Arvidsson R, Baumann H, Borg J, Keller E, Mila i Canals L, Selmer, J. K. A product chain organisation study of certified cocoa supply. 6th International Conference on Life Cycle Management, Göteborg, Sweden, 2013.
17. Camacho Otero J, Baumann H. Unravelling the shrimp nets.Tracing actors, arguments and life cycle thinking in the controversy over the sustainability of the Swedish West Coast shrimp. ESA report 2016:17, Chalmers University of Technology, Göteborg, Sweden, 2016.
18. Lindkvist M, Baumann H. Analyzing how governance of material efficiency affects the environmental performance of product flows: a comparison of product chain organization of Swedish and Dutch metal packaging flows. Recycling. 2017;2(4):23.
19. Venturini T. Diving in magma: how to explore controversies with actor-network theory. Public Underst Sci. 2010;19(3):258–73.
20. Latour B. Mapping controversies: syllabus 2012–13. MediaLab. Science Po. Retrieved from http://www.medialab-dev.sciences-po.fr, Accessed 15 Oct 2015.
21. Mitchell R, Parkins J. The challenge of developing social indicators for cumulative effects assessment and land use planning. Ecol Soc. 2011;16(2):29.
22. Bond A, Morrison-Saunders A. Challenges in determining the effectiveness of sustainability assessment. In: Bond A, Morrison-Saunders A, Howitt R, editors. Sustainability assessment: Pluralism, practice and progress. New York: Routledge; 2013. p. 37–50.
23. Franks DM, Brereton D, Moran CJ. Cumulative social impacts. In: Vanclay F, Esteves AM, editors. New directions in social impact assessment: Conceptual and methodological advances: Edward Elgar Publishing; 2011. p. 202–20.

Chapter 7
Sustainable Guar Initiative, Social Impact Characterization of an Integrated Sustainable Project

Marie Vuaillat, Alain Wathelet, and Paul Arsac

Abstract Sustainable Guar Initiative is a three-year long integrated program aiming at developing sustainable guar production within the Bikaner district in Rajasthan, India. SGI was set up by Solvay, L'Oréal, HiChem and the NGO TechnoServe. To confirm and consolidate the relevance of the program and to identify potential improvement opportunities, an environmental and social Life Cycle Assessment has been conducted, comparing the guar production before and after the Sustainable Guar Initiative. This paper focus on methodological aspects related to the Social Life Cycle Assessment (S-LCA): functional unit, scope and rating system definition. We set up a common rating system enabling information aggregation from multiple sources in order to capture the complexity of a product system.

7.1 Introduction

In 2013, Solvay and L'Oréal launched the Sustainable Guar Initiative (SGI) with a vision to leverage strong community-level partnerships in designing and implementing a program to promote socially and environmentally responsible practices in guar cultivation. Specifically, the program is expected to support the guar gum farmers in Rajasthan to adopt methods in accordance with the Food & Agricultural Organization's (FAO's) Good Agricultural Practices (GAP).

In order to design and implement actions in the Sustainable Guar Initiative, Solvay and L'Oréal partnered with TechnoServe to conduct an assessment of current activities, make recommendations and set-up a program. The program launched in 2015 is based on 4 themes: (1) Agronomy: enhancing sustainable practices for rain-

M. Vuaillat (✉)
EVEA, Lyon, France
e-mail: m.vuaillat@evea-conseil.com

A. Wathelet
Solvay, Brussels, Belgium

P. Arsac
L'Oréal, Clichy, France

© The Author(s) 2020

M. Traverso et al., *Perspectives on Social LCA*, SpringerBriefs in Environmental Science, https://doi.org/10.1007/978-3-030-01508-4_7

fed guar production, (2) Environment: groundwater-neutral approaches and best practices in guar farming, along with tree plantation, (3) Social impact: gender approaches, nutrition, health & hygiene and (4) Market improvement: traceability, supply chain and market access. The S-LCA aims at capturing improvement and threats linked with these themes but also with other themes if relevant.

7.2 Methodological Background

The S-LCA has been conducted according to already available guidance, including UNEP-SETAC Guidelines for Social Life Cycle Assessment of Products [1] and WBCSD Social Life Cycle Metrics for Chemical Products [2]. Diana Indrane's Master's thesis on "Integrating Smallholders within the Handbook for Product Social Impact Assessments" [3] has been a milestone in order to better take into account the smallholders specific issues.

7.3 Functional Unit

The subject of the study is guar seeds produced by program farmers in the Bikaner district, India.

Bikaner is a desert district situated in the North-West of Rajasthan, the largest Guar producing region in the state.

According to 2011 census [4], the economy of Bikaner district is mainly dependent on agriculture as 61.1% workers in the district are either cultivators or agricultural laborer's (meaning ~ 233, 787 households). There are currently 2274 households registered in the SGI program, representing around 0, 6% of the district's population and 1% of the district's agricultural laborer's. A vast majority belonging to the smallholder category (average landholding of 7.5 ha according to TNS last estimates).

Guar is sown in July and harvested in October. Thus, since the beginning of the program, three sowing and harvesting seasons have been conducted (2015, 2016 and 2017).

The functional unit has been defined as follow: « Producing 1 ton of Guar seeds in SGI program farms of Bikaner district, India »

The following scenario are compared:

- Before the SGI implementation, i.e. before May 2015
- Year I of SGI (May 2015–April 2016)
- Year II of SGI (March 2016–April 2017)
- Year III of SGI (March 2017–April 2018)

The main group of people involved in the realization of the function is farmers producing guar. But other group of people can beneficiate or be affected by actions and changes set up by the program: family members, neighbors, local community, society.

7.4 Scope

In order to select organization and group of people to evaluate, we followed the recommendations from WBCSD.

Main elements from discussion on organization's inclusion are described below:

- Fertilizer comes mainly from cow dung (owned by the same household growing guar), so no significant industry was identified at this step.
- The use of pesticides slightly increased due to the program, mainly as curative agent. According to surveys the maximum pesticides inputs is below 2% of the functional unit. An un-exhaustive list of pesticide was provided by Technoserve, among which dangerous reactive coming from large chemical producers. Thus, workers and local community of chemical plants could be exposed to dangerous reactive and products. Farmers and workers are not in direct contact with the pesticide producers but with resellers. Large chemical companies have high safety standards and L'Oréal and Solvay have few leverage on these companies' policies. Finally, the amount of pesticide products is still small compared to the other inputs.
- Seeds come ideally from the local University. Farmer producer group can purchase together seeds for economies of scale. It is part of the SGI to develop a safe supply chain with bulk purchase and storage (this part is still ongoing). No significant industry was identified at this step.
- Use of tractor is limited. 12 farmers out of the 150 questioned in the baseline survey own a permanent tractor. Rich farmers have tractor, 4 wheeler and other agriculture equipment required for farming, poor farmers have bullock cart/camel cart or hire tractor for carrying out agriculture operations. No significant industry was identified at this step.
- Supporting workers are from households who earn part of their livelihoods from seasonal wages. They are not migrant workers, but more neighbor agricultural workers. They have agricultural land but not enough to provide income for their family. Most of the supporting workers are from the local areas. For example, if the harvest has to begin in a village, then farmers may hire workers from another nearby village to conduct the harvesting operation (but also sometimes for weeding and hoeing).
- In general, there are 3–4 local traders in every village. Work is in progress with traceability and block chain to avoid sale intermediaries: commission agent, trader, broker currently often in place. Suppression of intermediaries would satisfy both Hichem (the griding company) and farmers. The job suppression

Table 7.1 Stakeholders inclusion

–	Solvay's influence level	Contribution to the functional unit	Potential risks
Pesticide suppliers's Workers and local communities	Low	Low	High
Smallholders	High	High	High
Local community	Medium	Low	High
Supporting workers	Medium	Small	High
Commission agent, trader, broker	Medium	Medium	Low
Hichem griding factory	High	Low	Low
Society	Low	Low	High

probability for commission agent, trader and broker was judged low by Technoserve as local traders have multiple portfolios including gram, groundnut, moth, sesame, etc. and because the volumes of SGI guar seeds represent extremely small quantities relative to total output in Bikaner.

- The griding company is Hichem (Hindustan Gum and Chemicals Limited, a 50% joint venture with Solvay) and is part of the program. Its activity and management is not expected to change due to the program.
- The SGI has no consequences on guar gum quality and thus on use and end of life steps.

As a summary, with 3 criteria: Solvay's influence level, contribution to the functional unit and potential risk (due to sectoral and geographical context), we can differentiate stakeholders that are directly or indirectly affected by guar farming. The SGI focuses on the principal directly affected stakeholders: smallholders. Because S-LCA is a systemic approach, we have to broaden the scope of analysis by also including indirectly affected stakeholders (Table 7.1).

There are potential overlaps between the areas of workers, local community and society (some people belong to smallholder, local community and society at the same time). However, assessing the three areas makes it possible to cover all potential risks, whereas assessing only the nearest perimeter does not guarantee that certain risk will be dealt with, particularly given the context of the duty of vigilance.

7.5 Social Topics

In order to select the relevant social topics, we followed 4 steps:

- Screening of PSIA Handbook [5] and WBCSD topics, selection of WBCSD mandatory topics (more numerous than PSIA's)
- Screening of SHDB hotspots for India and "crops" sector, selection of "High" and "Very high" tagged topics

- Addition of "project specific" topics on Smallholders and Society: topics covered by the SGI program or identified as relevant according to our expertise.
- Suppression of some topics due to irrelevancy

Three topics on workers were excluded because no organization, company or union is involved, so there is no management or policy in place on these topics.

- Freedom of association, collective bargaining and labor relations,
- Safety management system for workers
- Skills, knowledge and employability.

Working hours indicator was not included due to the specific nature of small-holder's activity. As they are their own bosses (self-employed, they define their own schedules). Working hours is rather a consequence of other aspects: sales price, profitability of the operation, access to basic services. . . .

Impact on consumer health and safety is also excluded because guar gum doesn't have specific positive or negative impact on consumer health and safety.

Guar farming household (smallholders) representing 60–80% of the total rural population (local community), we estimated that access to basic needs and health and safety was already covered by the same topics on smallholders.

"Job creation" was aggregated with "contribution to economic development" and "Respect for indigenous rights & delocalization" with "land titles" addressing the issue of land dispute (Table 7.2).

7.6 Defining a General Rating System

For some topics, included in the SGI program, monitoring and performance measurement is in place (e.g. women empowerment) but for other topics it is not the case (e.g. occupational health and safety of workers). In order to solve this issue, we collected new data and we developed a specific rating system enabling to deal with data heterogeneity among the social aspects, stakeholders and life cycle steps.

We started with the work from Technoserve based on the theory of change. This is based on the principle that activities are designed to address specific market failures/deficiencies (outputs), that will result in improved system performance measured by changed behavior (outcome), leading to greater financial benefits for targeted participants (goals).

The following elements are extracted from D. Indrane master thesis.

1. inputs are the resources necessary to carry out an activity,
2. activities are then implemented and effects (output, outcome, impact) can be analyzed,
3. output of the activities can be measured,
4. outcomes are the changes in the lives of the targeted population,
5. impact is an experienced improvement in lives of the targeted population.

Table 7.2 Social topics selection

	Directly affected stakeholders		Indirectly affected stakeholders	
Overarching social topics	Smallholders	Seasonal supporting workers	Workers and local community from pesticides companies	Society as a whole
Basic rights & needs	Meeting basic needs	Fair wages		Public commitment on sustainability
	Access to services and inputs	No child labor		Armed conflicts
	Women's empowerment	No forced labor, human trafficking and slavery		Corruption
	Child labor			
	Land titles			
Employment	Trading relationship			Contribution to economic development (including job creation)
	Next generation smallholders			
Health & Safety	Health and safety	Health and safety	Health and safety	
Skills & knowledge	Education and training			Contribution to technologic development and innovation

Outcome (4) and impact (5) level were left aside as it is currently difficult to disentangle the specific effects from inputs on outcome or impact (further down the impact pathway). So the ideal performance is considered as an output (3) from conducted inputs. In other words, the reference scale developed for smallholders is based on the first three steps of the Theory of change.

Focusing on the first three steps, we described them according to 2 criteria: implemented actions and fulfillment status (result).

- Actions can be the presence of monitoring, identification of opportunities, intervention, feedback monitoring.
- Fulfilment status can be legality or illegality, meeting the basic needs of a population fraction and positive feedback.

But no monitoring doesn't necessarily mean illegality or that basic needs are not met. These two criteria where not sufficient to describe all situations especially for some topics not included in the program thus, so not monitored. In order to be able to integrate unmonitored topics, we integrated risk assessment as another component of the rating system, enabling to fill the gap when no surveillance is implemented.

When using risk assessment, the question of the risk perimeter should be addressed. A sector and country risk does not necessarily mean a local risk. So generic data have to be finely tune with local information such as an individual testimony. Describing more precisely the testimony sources is then very important to evaluate its relevance. The amount of people testifying, the kind of person and its relation with the aimed stakeholder are important information to take into account. On the risk criteria, we defined the following inventory indicators:

- Country Specific Sector (CSS) risk based on SHDB
- Local risk and reported cases based on Technoserve testimony

We also experience the specific case of positive impacts. Positive impact can result directly or indirectly from an action. They rarely have a negative counterpart, but the question can still be addressed. It is therefore difficult to use risk assessment or theory of change for these aspects. The presence of positive or negative signals can be used. And so, the origin and type of signals can be used as rating criteria. For these topics, we were inspired by WBCSD rating system for this kind of topics (e.g. consumer's product experience).

- Weak signals
- General recognition
- Recognition by a specific study from the organization involved
- Recognition by a specific study from independent source

So the final rating system for our set of topics is detailed in the Table 7.3.

We used this general rating system for every social topic previously defined. But for some topics, further work was undertaken in order to capture the complexity of social phenomena (e.g. child labor).

7.7 Results

Major improvements are witnessed in smallholder's lives and activities thanks to the Sustainable Guar Initiative over the first 2 years of implementation. The third year is under evaluation. Areas of improvement were identified in order to insure long term improvement and more precise feedbacks (Fig. 7.1).

New areas of improvement were identified on indirect stakeholders. Mainly by setting surveillance and preventive actions.

Our work is an attempt to structure social impact assessment method. Risk, actions and results could be three main components of an integrated social impact assessment method enabling to aggregate the complexity and diversity of human aspects and heterogeneity of data available.

Table 7.3 General rating system

	Risk assessment	OR Result + Actions	OR Positive impact
+2	–	Basic needs met (>95%) OR legality	No negative signals, positive weak signals, positive impact generally recognized, specific study from company and independent actors recognizing positive impact
		AND intervention	
		AND positive feedback: Useful, informative and tailored (surveys)	
		AND continuous monitoring witnessing positive outputs and no deterioration	
+1	–	Basic needs met (>95%) OR legality	No negative signals, positive weak signals, positive impact generally recognized, specific study from company recognizing positive impact
		AND intervention	
		AND positive feedback: Useful, informative and tailored (surveys)	
0	No surveillance	Basic needs met (>95%) OR legality	No negative signals, positive weak signals, positive impact generally recognized
	AND risk in CSS is medium or low	AND intervention OR surveillance	
	AND local risk low (independent source)		
−1	No surveillance	Basic needs not met OR illegality	No negative signals, positive weak signals
	AND risk in CSS is high/very high AND local risk low (independent source)	AND intervention OR surveillance	
−2	No surveillance	Basic needs not met OR illegality	Negative signals, no positive weak signals
	AND risk in CSS is high/very high AND cases reported/ local risk is high (independent source)	AND no surveillance	

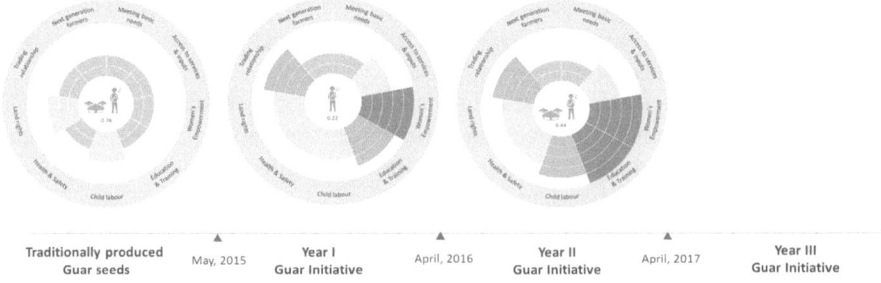

Fig. 7.1 S-LCA results on Smallholders, until year II

References

1. UNEP/SETAC. Guidelines for social life cycle assessment of products, 2009.
2. WBCSD. Social life cycle metrics for chemical products, 2016.
3. Indrane D. Small but complex: integrating smallholders within the handbook for product social impact assessments, 2017.
4. Directorate of census operations Rajasthan. district census handbook, 2011.
5. Fontes J. Handbook-for-Product-Social-Impact-Assessment-30, 2016, 1–146.

Chapter 8
Generation, Calculation and Interpretation of Social Impacts with the Social Analysis of SEEbalance®

Peter Saling, Ana Alba Perez, Peter Kölsch, and Thomas Grünenwald

Abstract Measuring sustainability is an important prerequisite for making strategic decisions. BASF has developed several instruments to evaluate sustainability whereby the utilization of each method depends on the concrete purpose or issue in question. The new Social Analysis will contribute to this setup by assessing social impacts along the value chain.

The Social Analysis is implemented in the SEEbalance® calculating results with the Social Life Cycle Assessment (S-LCA) and with a specific Social Hot Spot Assessment. Both approaches generate, calculate and interpret the social impacts from different perspectives. Different levels and approaches of data generation and calculation are used for drawing conclusions on the social performance of product alternatives fulfilling the same functional unit. The close link to the environmental life cycle assessment enables practitioners a holistic view on sustainability aspects which in turn support decision-making processes.

For the assessments, processes and decision trees to harmonize the generation of coherent results were developed and will support the data generation process. Different levels of interpretation of findings leading to overall results support and harmonize the interpretation of the findings significantly.

Keywords Social Life Cycle Assessment (S-LCA) · Social Hot Spot Analysis · Interpretation of results · Sustainable Development Goals (SDGs) · Social impacts

8.1 Introduction

The quality and quantity of data on appropriate social indicators have often been insufficient. This is particularly relevant for those requiring a relatively high degree of in-depth details. While assessing product alternatives for a defined functional unit, the scope of the needed data for a social assessment differs from those for an

P. Saling (✉) · A. A. Perez · P. Kölsch · T. Grünenwald
BASF SE, Sustainability Strategy, Ludwigshafen, Germany
e-mail: peter.saling@basf.com

© The Author(s) 2020
M. Traverso et al., *Perspectives on Social LCA*, SpringerBriefs in Environmental Science, https://doi.org/10.1007/978-3-030-01508-4_8

Table 8.1 Different approaches for environmental LCA and S-LCA

Data	LCA	S-LCA
Availability in a meaningful approach		Coutry specific
		Sector specific
	Company specific	Company specific
	Site specific	
	Product specific	
	Process specific	
Meaningful level of quantification		Qualitative
		Semi-Quantitative
	Quantitative	Quantitative

environmental evaluation. Two processes that deliver the same product might vary in the results of the Carbon footprint or Water footprint. Due to the corresponding technologies, an in-depth life cycle assessment (LCA) on the environmental performance is advisable.

On social aspects this approach may most commonly does not lead to conclusive results. The social impacts of two products produced in the same factory or company assessed on specific indicators will frequently generate uniform results. Assessing the indicators 'wages and salaries of the workers' or 'working hours' in a factory or company will show a similar dimension for both products. Hence, an analysis on this level might generate artificial or irrelevant discrepancies. A plant manager would not benefit from this approach and the identification of social improvement potentials is lost. Consequently, social data need to be handled and interpreted with other approaches (see Table 8.1).

SEEbalance® integrates the Eco-Efficiency Analysis [1] that evaluates environmental factors and costs of a product with the Social Analysis assessing social indicators to a holistic picture which allows practitioners the interpretation of all relevant information in an overview system. The first version was developed in the year 2005 using a different methodology [2–6].

8.2 Methods

Coherent databases are vital for social assessments of each step of the life cycle. The goal is the combination of technical expertise, know-how on supply chains and the link to meaningful social indicator assessments. The selection of appropriate indicators and the aggregation into a result are key aspects in social assessments. Therefore, the SEEbalance® can be executed on different levels of information and with different views on a supply chain to identify the relevant disparities between the evaluated alternatives. The approach consists of four modules that together address economic, environmental and social impacts. The Social Analysis

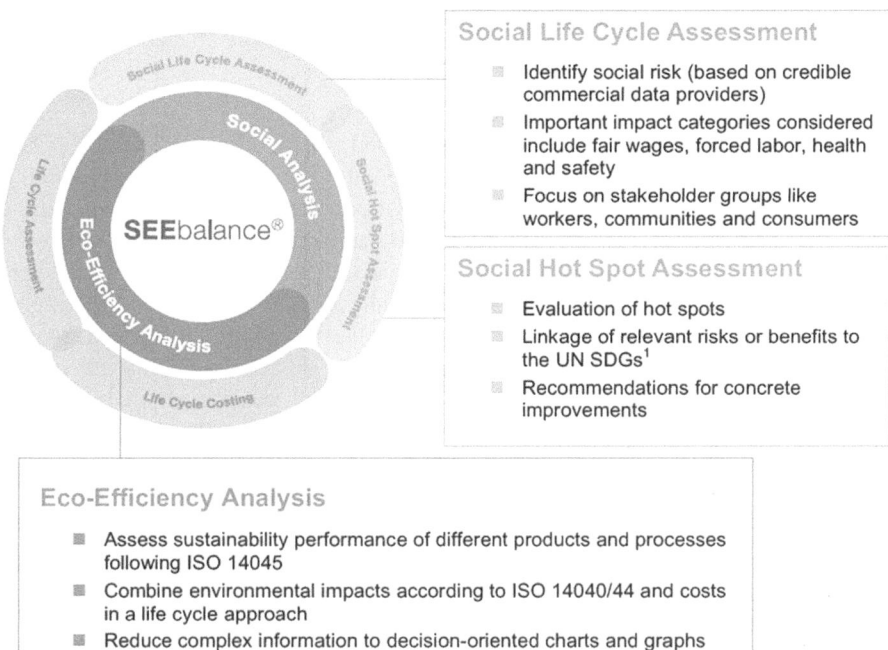

Fig. 8.1 **SEE**balance® is comprehensive approach to assess social, environmental and economic aspects over the whole life cycle of product or process alternatives

covers the assessment of social impacts and consists of two modules, namely the Social LCA and the Social Hot Spot Assessment (see Fig. 8.1).

8.2.1   Social LCA

The New SEEbalance® considers economic, environmental and social aspects. The **Social Analysis** for social aspects and here specifically the S-LCA takes the following new developments into account:

- Roundtable for Product Social Metrics: 3 stakeholder groups, 19 social topics [7]
- WBCSD – Social metrics for chemical products: 3 stakeholder groups, 25 social topics (11 mandatory), 168 indicators [8]
- UNEP/SETAC: 5 Stakeholder categories, 6 Impact categories, 31 Subcategories [9]

8.2.1.1   The Handbook of the Roundtable for Product Social Metrics

The handbook provides an overview of indicators, social topics and assessments. It considers the UNEP SETAC Guidelines for Social Life Cycle Assessment of Products [9] and corporate level standards (GRI, 2013; ISO 26000:2010). Given the lack of global standards on methods for social impact assessment at the product level, the Roundtable developed this method through gaining an understanding of and drawing upon the various methods already applied by their members [10].

The guidelines proposed in the handbook uses 12 key principles. The principles provide both the guiding rules that were considered during the handbook development and the foundation on which companies can assess product social impacts. Such a principle described in the Handbook is e.g. that as a guideline for product social sustainability, one should focus on the practical feasibility for companies to use and implement the method within their respective organizations, allowing businesses to develop it organically, as well as to improve performance based on an aligned and transparent method.

The principle for defining an impact is outlined as follows: social topics and performance indicators should reflect positive and negative impacts of the product to enable a reasoned assessment of the overall performance (adapted from the balance principle of the GRI).

Furthermore, the assessment should cover social topics that are significant for the overall evaluation of the social impact of the product and which may have an impact on the business and/or influence external stakeholders' perceptions of the product (principle for identifying relevance, adapted from the materiality principle of GRI).

Data used to support the assessment should be gathered, recorded, compiled, and in the event of external verification and eventually be disclosed in a way that preserves the quality and the relevance of the information (principles for data and verification, adapted from the reliability principle of GRI). During the Social Analysis development, these principles were mainly applied for the indicators, the social topics and the assessment of the social topics including its interpretation and reporting.

8.2.1.2   WBCSD – Social Metrics for Chemical Products

The development of the guidance follows six principles: relevance, completeness, consistency, transparency, accuracy, and feasibility.

The work and the generation of the guidance document has been mainly inspired by the Guidelines for Social Life Cycle Assessment of Products and associated works (UNEP/SETAC, 2009), and the 2014 Handbook for Product Social Impact Assessment, v. 2.0, developed by the Roundtable for Product Social Metrics. Additionally, investigations among stakeholders on social impacts and previous work performed with the WBCSD on life cycle metrics devoted to environmental impact assessment were also used as a base.

25 social topics were selected as the most representative and were split into two groups: 11 mandatory social topics need to be covered as a minimum in every product assessment. Indicators and optional advanced indicators were proposed for each mandatory and non-mandatory social topic. 14 non-mandatory social topics may be included in a product assessment based on a selection process defined in the present guidance. Indicators and optional advanced indicators are also proposed for each non-mandatory social topic. Three stakeholder groups are targeted for this work on social metrics: workers, consumers and local communities.

In both frameworks, five reference scale levels enable the valuation of each indicator and advanced indicator: from −2 (unacceptable performance) to +2 (outstanding/exemplary evidence), via 0 (standard performance/compliance).

8.2.1.3   Social LCA Development

For the development of the Social LCA within SEEbalance® impact categories and their indicators were carefully assessed, following the principles and guidelines of these frameworks and publications. The key aspect in their selection was their coverage in different databases and their availability. Ultimately, the method aggregates and summarizes different levels of data. The indicators and their underlying definitions were chosen by evaluating the best fit and practicability. Table 8.2 shows the overview of impact categories used in the S-LCA and the indicators linked to the impact categories.

For example: freedom of association is an important indicator for workers and can be measured in the different systems with the indicators "Fundamental Human Rights (Social)" from Ecovadis™ [11], "Freedom of Association and Collective Bargaining" from Reprisk® [12] and "Freedom of Association and Collective Bargaining Index" from Verisk Maplecroft™ [13]. The latter database is used if no company information is available. They are based on country level data or mixtures of data on the country level which are generated and collected in a specific database.

8.2.2   System Boundaries

Based on the environmental LCA, that applies the definition of system boundaries integrating all relevant life cycle steps, the social assessment should follow the same approach.

The clear distinction of all life cycle steps is important as the number of single results per life cycle step influences the result. In the environmental LCA where e.g. Carbon footprints or other environmental impacts can be mathematically aggregated along the life cycle, the results of the calculation show which inputs have relevant contributions through their numeric information. The cut-off criteria can be derived and used to limit the number of life cycle steps that should be considered.

Table 8.2 Mapping of selected impact categories with data sources indices

Stakeholder group	Impact category	Ecovadis™ (selected social criteria)	Reprisk®	Verisk maplecroft™
Workers	Health & safety	Employee Health and Safety (Social)	Occupational Health and Safety Issues	Occupational Health and Safety
	Fair wages	Working Conditions (Social)	Poor Employment Conditions	Decent Wages Index
	No child labour	Child & Forced Labour (Social)	Child Labour	Child Labour
	Appropriate working hours	Working Conditions (Social)	Poor Employment Conditions	Decent Working Time Index
	No forced labour	Child & Forced Labour (Social)	Forced Labour	Forced Labour
	Freedom of association	Fundamental Human Rights (Social)	Freedom of Association and Collective Bargaining	Freedom of Association and Collective Bargaining Index
	No discrimination	Discrimination & Harassment (Social)	Discrimination in Employment	Discrimination in the Workplace
Local communities	Healthy and safe living conditions		Impacts on Communities	Healthcare Capacity Index
	Security and conflict		Human Rights Abuses, Corporate Complicity	Security Forces and Human Rights Index
	Land and property rights		Local Participation Issues	Land, Property and Housing Rights Index
Consumers	Healthy and safe products	Customer Safety (Environment)	Products (Health and Environmental Issues)	Efficacy of the Regulatory System

For the assessment of social aspects, we have experienced that mass flows and cut-offs, as used for environmental LCA, often do not necessarily reflect the relevant impacts. Small mass flows can be as relevant as higher mass flows. A reduction of a Carbon footprint by 20 % means a significant reduction and a significant figure. The use of steam, electricity etc. can be assessed quite well and the corresponding Carbon footprint can be calculated with a low uncertainty.

For social indicators the direct link to mass flows does often not commonly lead to conclusive results, especially for figures which are based on statistical data. For example: working accidents are derived for a whole company using a statistical number of working hours as the basis. Theoretically, the number of accidents can be allocated to specific sites or processes via working hours linked to them. However, in reality this does not reflect the accidents in a specific factory for a specific number of products. A comparison of a process with higher working hours is not necessarily linked to a higher rate of accidents that were calculated statistically. Another example is child labour: for instance, if a process is related to child labour, a product comparison might end in comparing 1 h of child labour with 2 h of child labour per kg of the final product depending on the amount of materials needed respectively. This however does not inevitably mean that one product is better than the other as both are affected by a severe social issue. In such a case it would be more important to detect if child labour is involved or not, which means the need of a more general and qualitative approach that allows the identification of product related impacts in a more meaningful manner. Likewise, mass flows are not relevant as even a low material input can be significant as could be seen in the case of child labour.

Previous studies conducted with the SEEbalance® approach showed that the material amounts as a basis for the assessment of social indicators generate mislead-ing results in some situations. Therefore, for the new Social Analysis within SEEbalance® we decided to shift from a mass-based and quantitative approach to a qualitative approach. The basic intention was to identify risks and risk potentials of products along the whole value chain. Therefore, the supply chain is analyzed accurately to identify life cycle steps with relevant social risks after the system boundary sheet has been created.

8.2.3   Assessments of Companies and Countries

Depending on the level of available information, either social impacts of companies or of countries are considered in the Social Analysis. A hierarchy of data sources is applied to use the most accurate levels of information. Starting with the highest level company data are considered, followed by the assessment of countries and mixes of countries. A decision tree was created starting with a company specific assessment using Ecovadis, followed by a Reprisk research and subsequently followed by a country/sector specific analysis using Maplecroft data (see Fig. 8.2).

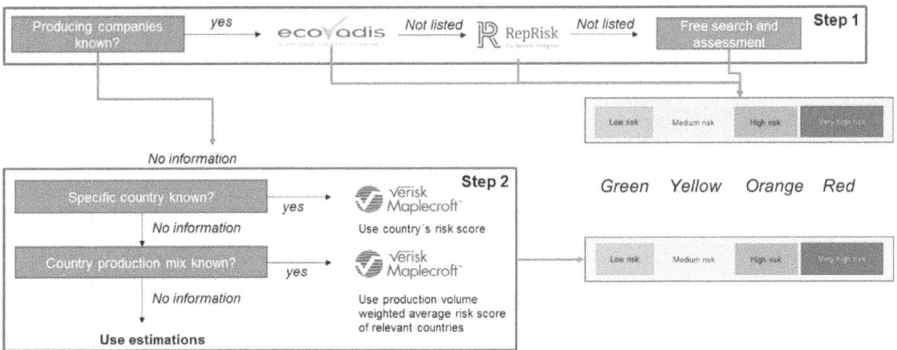

Fig. 8.2 Two step approach depending on data availability for the Social LCA

8.2.3.1 Company Specific Assessments

A company specific assessment is prepared based on the databases Ecovadis™ or Reprisk®. As Ecovadis is based on an agreed process and primary data out of a review process, this information is the first choice if it is available in the specific case. The results of the assessment are transferred to a scale based rating from 1 to 10 with each of them linked to a colour code for the ranking from green to red in four steps. If no data on the company level are available in Ecovadis, a Reprisk research on company specific social risks will be performed. The information that are identified will be verified and linked to the indicator list of the S-LCA. Following a decision tree, one decides if an incident is confirmed or not as well as if transparent management processes are in place (Fig. 8.3).

Additionally, further options, statements and published programs of the companies are checked and transferred to the rating system ending in a specific colour code for the respective company.

Definition of a transparent management process (based on the "code of conduct" of BASF)

- Governance and **policy**
- A position of the company for the respect and support of human rights is published and it is announced as mandatory for the company and its employees to act in accordance with internationally declared human rights and adhere to applicable laws within the framework of the business activities.
- **Implementation** in the company
- The company describes, how it ensures that the positions and guidelines are implemented in transparent targets, processes, management systems and

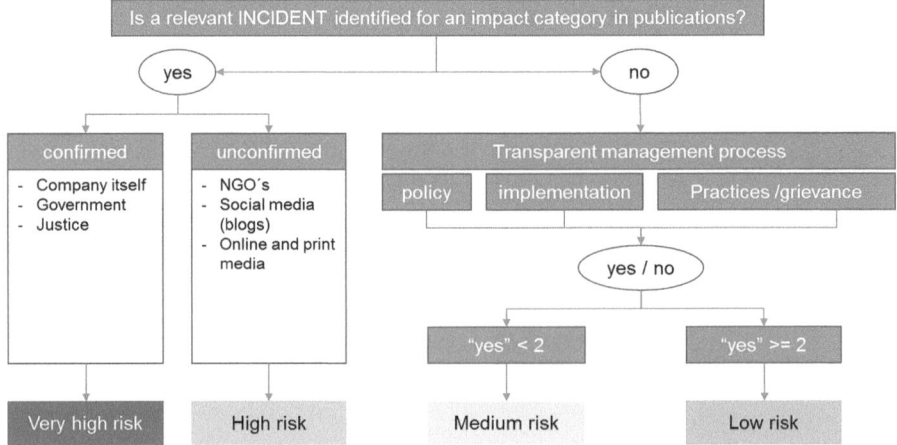

Fig. 8.3 Assignment of an incident to the risk categories, assessment of transparency of management processes

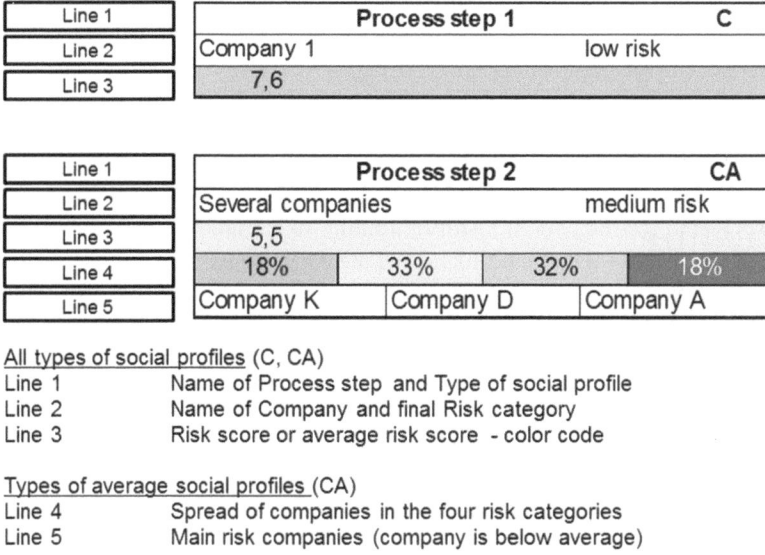

Line 1	**Process step 1**		C
Line 2	Company 1	low risk	
Line 3	7,6		

Line 1	**Process step 2**			CA
Line 2	Several companies		medium risk	
Line 3	5,5			
Line 4	18%	33%	32%	18%
Line 5	Company K	Company D	Company A	

<u>All types of social profiles</u> (C, CA)
Line 1   Name of Process step and Type of social profile
Line 2   Name of Company and final Risk category
Line 3   Risk score or average risk score - color code

<u>Types of average social profiles</u> (CA)
Line 4   Spread of companies in the four risk categories
Line 5   Main risk companies (company is below average)

Fig. 8.4 Assessment results of social profiles for single companies and averages of companies

monitoring systems. A reporting and auditing process show the progress and developments of the basic requirements.

- Performance **practices,** remedies and **grievance** mechanism

The company shows and explains which measures were applied to improve unsatisfactory situations or performances on social standards continuously over a longer time period of several years. It is described, how the company accurately evaluates, whether it follows to the internationally recognized labour and social standards stated in the position and how violations against the positions are identified and subsequent remedies are initiated.

If more than one company delivers a specific product an aggregated figure of the contributions of all companies in a mass-allocated approach is generated as a result for this life cycle step. The results are collected for each life cycle step in company specific social profiles named "C" and averages of social profiles ending in aggregated social profiles named "CA". This social profile "CA" also shows a range of companies contributing to this social profile, the spread of the companies and an overview of all contributions that are observed for the four different social risk categories (see Fig. 8.4).

8.2.3.2   Country Specific Assessments

If no information on the company level is available as explained in Fig. 8.2, country related information will be assessed. The database used for this step is Maplecroft, issuing the selected indicator results for the Social Analysis. The new version of this database delivers sector-specific information for different countries which enables practitioners to distinguish social information for different sectors. For a specific country, the defined sector results can be extracted from the database in a 0–10 scale. Additionally, a weighting step is included to avoid the underestimation of impacts with a high severity. Illustrated with the impact categories "child labour" and "fair wages" it seems rational that child labour is more severe for the stakeholder group "workers" and for the overall result of a rating of a country than "fair wages".

Figure 8.5 shows the transformation of the database information to the result for a life cycle step. For the weighting scheme factors from a publication of the "Institut für Entwicklung und Frieden" [14] were applied.

In the same manner country mixes for different precursors were no single country can be identified, production statistics are applied including mass allocation.

The production as supplier volume in combination with the assessment factor generate an overall result for this life cycle step (Table 8.3).

In the next step the results are displayed to get an overview of the assessment of each life cycle step. The countries which are significantly contributing to the result of the life cycle step are highlighted. The spread of the contributions to the four risk categories, the total number and the colour code of the assessment are summarized and shown for every single life cycle step (Fig. 8.6).

8.2.4   Results and Interpretation

The results of the evaluation of different life cycle steps can be ordered and differentiated by country, country mixes, company and company mixes. They can

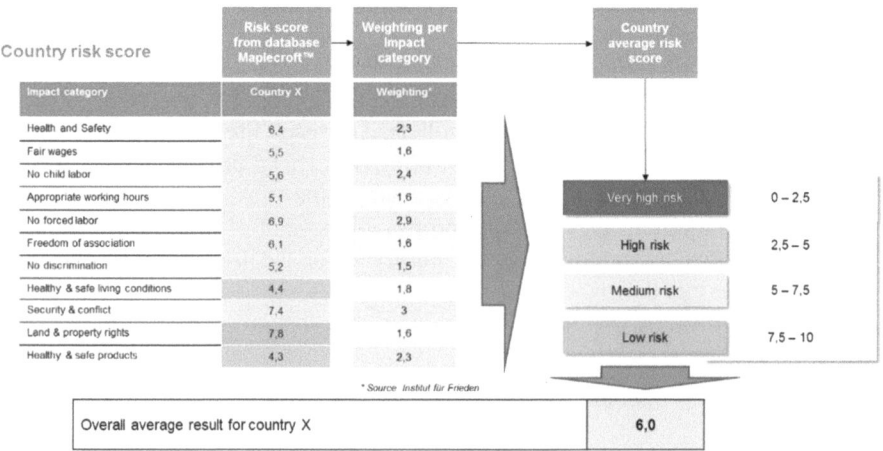

Fig. 8.5 Assessment results of social profiles for single countries

Table 8.3 Mapping of selected impact categories with data sources indices

Country	Production volume	Share	Rank	Assessment
Gabon	1.800,0	9,9%	4	3,0
Australia	3.000,0	16,5%	2	8,3
Brazil	1.000,0	5,5%	5	3,6
India	950,0	5,2%	6	3,0
China	2.900,0	16,0%	3	2,6
Malaysia	400,0	2,2%	8	4,1
Ukraine	390,0	2,1%	9	4,1
Kazakhstan	390,0	2,1%	9	4,6
Ghana	390,0	2,1%	9	3,6
Mexico	240,0	1,3%	12	4,1
Myanmar	100,0	0,6%	13	2,0
South Africa	6.200,0	34,1%	1	5,4

		high risk	Average risk score - color code
	Manganese world	4,7	
Spread contributions to the four risk categories	17% 34%	47% 1%	
Name of Process step and Type of social profile	Manganese production world	RA	
Spread of countries in the four risk categories	China, Gabon, India		

Fig. 8.6 Assessments of social profiles for averages of countries

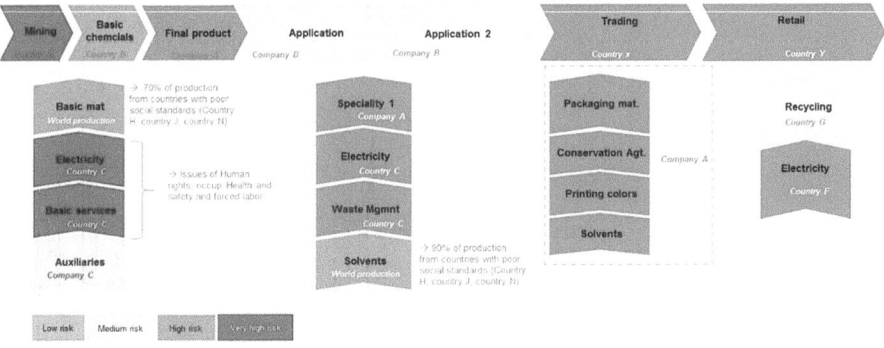

Fig. 8.7 Life cycle steps after its assessment in four different colour codes

be handled as an input data information as well as classified and displayed to achieve elevated levels of transparency for the result interpretation (see Fig. 8.7). Furthermore, the level of detail is relatively high and allows a deep dive into each process step for further investigation.

The results of the assessment are expressed in a specific four folded colour code system from "red = very high risk" via "orange = high risk" and "yellow = medium risk" to "green = low risk". As outlined in Table 8.1, it is more meaningful to express results in a qualitative or semi-quantitative manner to generate a better understanding of the social impacts along a value chain linked with a defined functional unit.

Following this approach, based on company and country sector level assessments, a translation of the findings in a 0–10 scale is the first step towards the generation of the specific colour for the scale-based approach. The whole system boundary sheet is transferred to displaying every single life cycle step with the corresponding colour codes in an overall view. This allows an easy identification of improvement potentials or opportunities for changing the process. For several alternatives that can fulfill the same functional unit, this approach allows a direct comparison of alternatives. It also generates a meaningful and transparent platform for decision-making processes. In the following an example is visualized in Fig. 8.7.

8.2.5 Interpretation of Results from the Life Cycle Assessment Based on Graphical Assessments

Different LCA methods require different approaches for the interpretation of the assessed data. A meaningful interpretation is linked with the application of specific instruments expressing results from the assessment in an easy and understandable way. The LCA and costs evaluation can work with specific figures which are aggregated by normalization and weighting. For the S-LCA a more qualitative approach is much more helpful because normalization and aggregation does often not lead to meaningful results. It was decided for the S-LCA to use a ranking approach based on several steps of the interpretation (see Fig. 8.8).

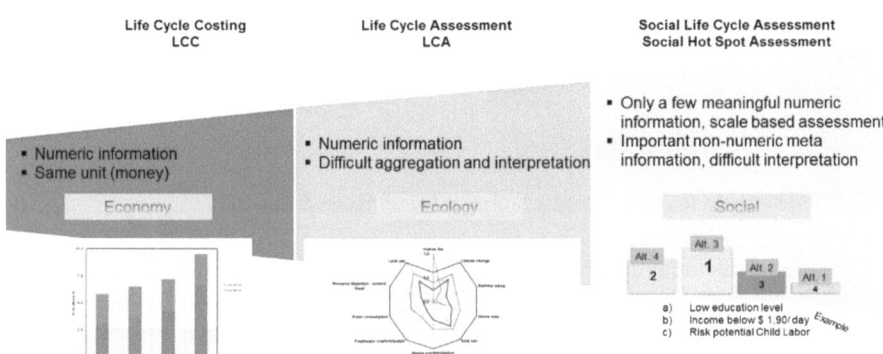

Fig. 8.8 Different approaches and challenges for assessing costs, environment and social impacts by using life cycle thinking

8.2.6   Interpretation of Results from the Life Cycle Assessment Based on Graphical Assessments

The collection and visualization of the life cycle steps allows a comparison of results of alternatives by applying a four step approach, combining mathematical and graphical elements. Fig. 8.9 shows a graphical element and how different life cycle steps and their social profiles can be displayed.

8.2.7   Interpretation of Results from the Life Cycle Assessment Based on Numerical Assessments

The numeric assessment of a single value is another approach for the interpretation of results from a S-LCA. The basic assessment of results based on a 0–10 scale linked with four different risk characterization classes also allows the averaging of figures and the summarization of different risk classes. In the overview alternatives with different numbers of life cycle steps linked to the risk categories can be compared.

If an alternative has much more life cycle steps with very high or high risks compared to an alternative which has no impacts in the "very high risk" classification and higher numbers of medium or low risk steps, an average of this alternative can be identified and used in the interpretation of results (see Fig. 8.10).

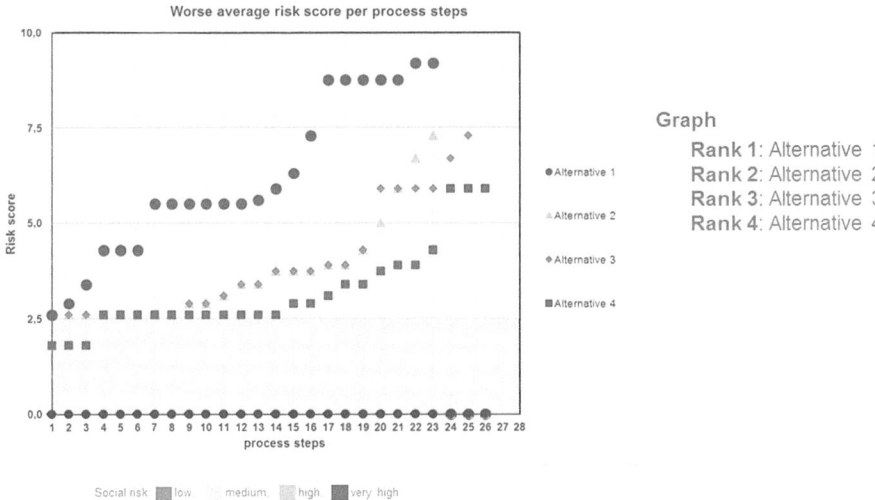

Fig. 8.9 Interpretation of social impacts in different life cycle steps. Graphical comparison of alternatives as basis for decision-making process

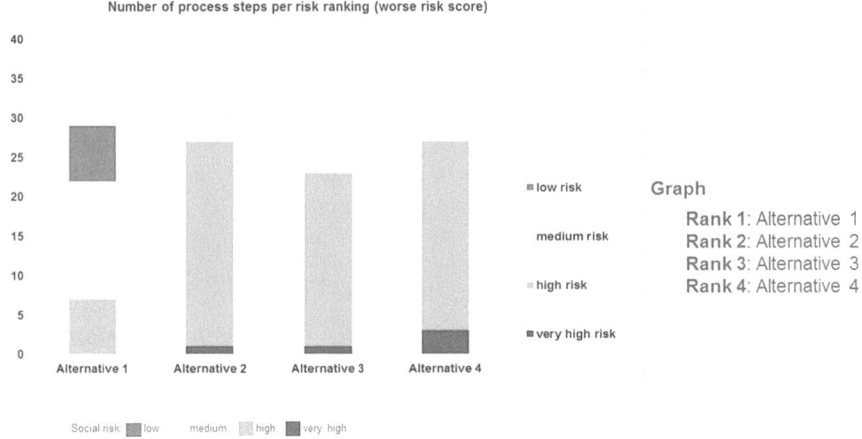

Fig. 8.10 Interpretation of social impacts in different life cycle steps. Numeric comparison of alternatives as basis for decision-making process

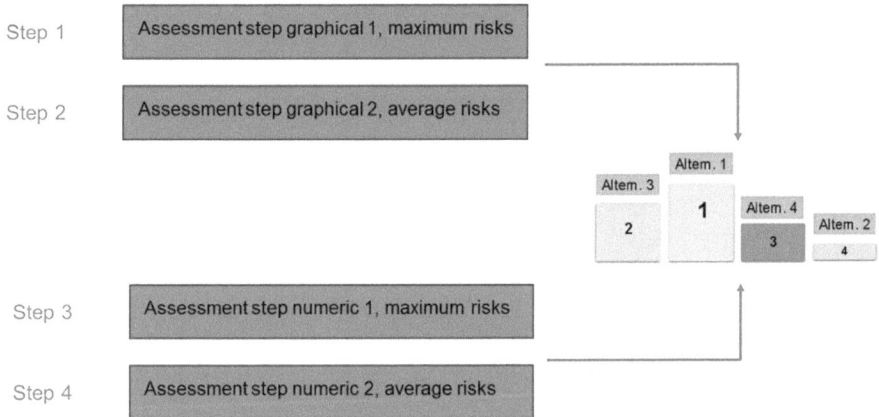

Fig. 8.11 Interpretation of all social impacts based on a four-step approach to a final ranking of alternatives

With the help of the graphical elements final conclusions can be generated, leading to a ranking of the alternatives (Fig. 8.11).

8.3 Social Hotspot Assessment

Social hot spots are assessed in the Social Hot Spot Assessment which contributes together with the S-LCA to the overall result of the Social Analysis. Meaningful social hot spots along the corresponding value chain are identified and evaluated.

The focus of this assessment step can be summarized as

- Deep dive into social hot spot(s) of the value chain
- Expert evaluation of relevant topics considering the Sustainable Development Goals (SDG)
- Identification of main social focus topics discussed by stakeholders
- Product-industry and region-specific analysis of social hot spots

8.3.1 Identification of Life Cycle Steps for a Social Hot Spot Assessment

In the assessment process of the whole supply chain the most significant life cycle step(s) is/are identified and highlighted. It is independent from the mass of materials that is needed to fulfill a functional unit as, as already applied in the S-LCA, even small volumes of materials can cause high impacts on social aspects. In contrast to the S-LCA, a limited number of life cycle steps is assessed, and it is therefore important to identify the most relevant steps. For each significant step, a deep dive assessment is conducted during which one will search for all information on social issues using different sources.

The goal of this step is the identification of hot spots beyond statistical figures that might negatively affect the marketing of the product in the defined application. The identification of the relevant life cycle steps follows the logic of determining each identified unique and specific steps that have a high importance to social indicators. For the assessment of alternatives, the perception of specific life cycle steps by the public and different groups of stakeholders should especially be considered and subsequently analyzed. Furthermore, it is important to focus on life cycle steps which are linked to a high risk.

It is helpful to follow the applicable guidance for the identification of the most relevant life cycle steps:

- Most salient social risks
- Need for more detailed information and contextualization
- Actual and potential impacts in own operations and through business relationships
- Link to social risk upon: direct/indirect business relationship, leverage potential, transparency on actors, etc.
- Relevance for stakeholders
- Recent and evolving public attention

For example: the T-Shirt production life cycle has a lot of single life cycle steps. Significant steps can be working conditions in some countries or companies producing the shirts. The crude oil production in other countries as precursor for textile chemicals might be of less importance in this example.

8.3.2 Social Hot Spot Assessment Procedure

The Social Hot Spot Assessment follows a clear procedure to ensure that the process is coherent and delivers meaningful and reproducible results. The SDG definitions can be utilized after a free search to structure the search process and to enable linkages to the SDGs afterwards.

The process steps can be defined as:

- Starting with a free desktop research on all social topics that can be found in internet, literature, social media, etc.
- Filter the relevant information and reduce the information to a significant expression of the finding.
- Cross-check in a detailed analysis if there are social impacts that were not considered so far. Use the SDG goals and the sub-goals as guiding principles (without the environmental goals).

Step 1: Sorting of the Findings by Using Key Words

- Using the descriptions and sub-goals as orientation
- Select information as close as possible to the specific region/industry in the assessment
- Use official independent sources and sources from NGO or local sources
- Assess all SDG and try to find information on them
- If no information from general search can be found, search specifically for key words of SDG to complete the assessment
- Use most recent sources and information, do not consider information older than 10 years

Step 2 – Linking to SDG Phrase the findings by extracting from the searches

- Using meaningful phrases describing the finding accurately and link them via key words with SDG
- If one main category of SDG is found it will be counted as an impact
- If there are contradicting information available, do not link the information to a SDG
- Link the findings to SDG with short statements or explanations and generate an overview on them.

Step 3 – Summarize List all identified SDG in a graphical overview

- Make expert statements (up to 4 main social hot spots) in a nutshell
- Discuss improvement potentials
- List positive effects of the application of the product in focus to SDG as well in a written comment. Leave out SDG that are not relevant or where no information was found.

8.3.3 Social Hot Spot Assessment Interpretation

Key findings from the Social Hot Spot Assessment are translated into key words which are defined in the SDGs. During the matching process a SDG is identified if an alternative of the study shows significant assignable social issues. It will be identified and highlighted if and how life cycle activities and actors conflict with the respective SDG. In Table 8.4 it is shown how search results can be listed to the SDG descriptions and how findings from intense desktop researches, audits or other information can be used to identify activities working against specific SDGs. If a

Table 8.4 Search functions and linkage to identify incidents where SDG are negatively affected

SDG	Description	Finding	Result
5.1	End all forms of discrimination against all women and girls everywhere	In this sector women are much less paid than men	x
5.2	Eliminate all forms of violence against all women and girls in the public and private spheres, including trafficking and sexual and other types of exploitation	Violence against woman was reported in several cases	x
5.3	Eliminate all harmful practices, such as child, early and forced marriage and female genital mutilation	No indication	
5.4	Recognize and value unpaid care and domestic work through the provision of public services, infrastructure and social protection policies and the promotion of shared responsibility within the household and the family as nationally appropriate	No indication	
5.5	Ensure women's full and effective participation and equal opportunities for leadership at all levels of decision-making in political, economic and public life	Women are not allowed to work on management level of companies, no single female in a leading level of companies were identified in this region/sector	x
5.6	Ensure universal access to sexual and reproductive health and reproductive rights as agreed in accordance with the Programme of Action of the International Conference on Population and Development and the Beijing Platform for Action and the outcome documents of their review conferences	No indication	
5.a	Undertake reforms to give women equal rights to economic resources, as well as access to ownership and control over land and other forms of property, financial services, inheritance and natural resources, in accordance with national laws	No indication	

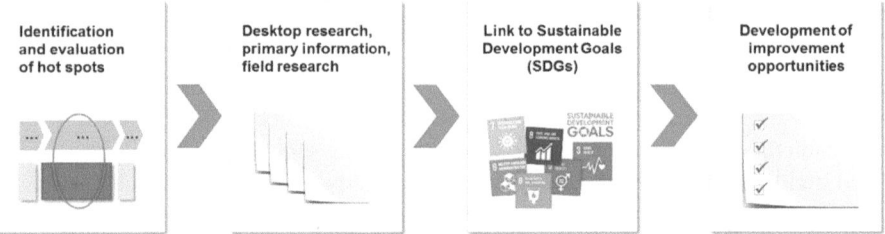

Fig. 8.12 Identification and assessment of possible Hotspots in the value chain and assessment of the Hotspots using the SDG descriptions

SDG group is identified where activities in the market, the region or the industry impairs the goals of SDG, it can be considered during the decision-making processes and the overall results of a SEEbalance®.

The method focusses on a deep dive into selected life cycle steps and assesses different types of risks. The SDGs support the identification of Hot Spots with its definitions and goal descriptions. It will be measured if and how life cycle activities and actors conflict with the SDG and subsequently summarized in an overview figure (Fig. 8.12).

8.4 Conclusions

The Social Analysis implemented in the SEEbalance® calculates results from S-LCA and from the Social Hot Spot Assessment. Both approaches generate, calculate and interpret the social impacts from different angles along the whole value chain of products or processes. The different modules can be applied separately from each other but can as well be integrated in an overall sustainability assessment result, also considering the results from the Eco-Efficiency Assessment (Fig. 8.13).

In summary, all three dimensions of sustainability can be covered in the SEEbalance® approach and give a holistic view for decision-making processes. SEEbalance® supports decision makers in identifying the sustainability benefits and trade-offs along the value chain of products or processes (Fig. 8.14).

8.5 Future Developments

The S-LCA will be challenged in defining relevant data sources, a better interpretation of data and a more detailed analysis of the combination of country and sector information. Different software approaches will be tested and to figure out how company specific data can be generated more efficiently.

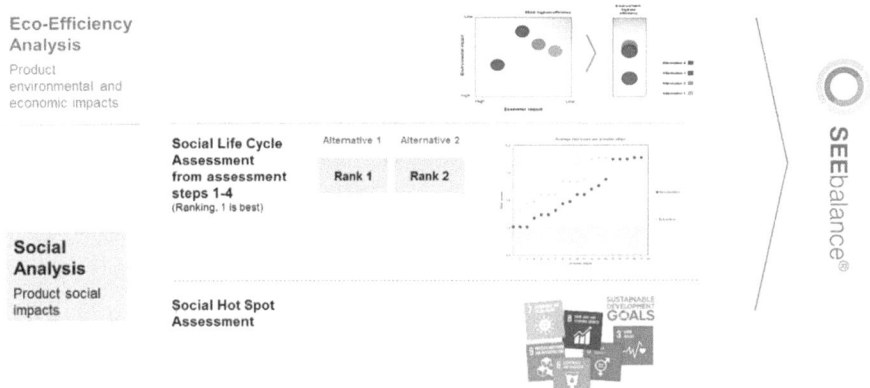

Fig. 8.13 Overview of economic, ecological and social product performances with the SEEbalance®

Fig. 8.14 Overview the main application fields of the SEEbalance®, respectively of single modules of it, depending on the question to be answered

The Roundtable for Social metrics will support these developments with BASF contributing by finding more opportunities for data generation and application.

The Hot Spot Analysis will demonstrate how different information can be transferred to a meaningful result. Furthermore, the last focus lies on how can the SDGs be used to assess and integrate them more effectively in the analysis to ensure:

- Deep dive into social hot spot(s) of the value chain
- Expert evaluation of relevant topics considering the SDGs
- Identification of main social focus topics discussed by stakeholders

References

1. Saling P, Kicherer A, Dittrich-Kraemer B, Wittlinger R, Zombik W, Schmidt I, Schrott W, Schmidt S. Eco-efficiency analysis by BASF – The method. Int J LCA. 2002;7(4):203–18.
2. Schmidt I, Meurer M, Saling P. Kicherer A, Reuter W, Gensch CO. SEEbalance – managing sustainability of products and processes with the socio-eco-efficiency analysis by BASF, Greener Management International, Greenleaf publishing Sheffield, S. Seuring (guest editor), Issue 45, Spring 2004, 79–94.
3. Kölsch D, Saling P, Kicherer A, Grosse-Sommer A. How to measure social impacts? What is the SEEbalance® about? – socio-eco-efficiency analysis: the method. Int J Sustain Dev. 2008;11 (1):1–23.
4. Saling P, Grosse-Sommer A, Alba-Perez A, Kalisch D. Using the eco-efficiency analysis and SEEbalance in the sustainability assessment of products and processes. In: Sustainable neighbourhood, from Lisbon to Leipzig through research, 4th BMBF-forum for sustainability, Leipzig, Germany, May 2007, pp 8–10.
5. Saling P, Pierobon M. Measuring the sustainability of products: The eco-efficiency and SEEbalance® analysis, LCM 2011, Berlin, http://www.lcm2011.org/papers.html, 21.11.2011.
6. Krueger C, Saling P. Chemistry powers sustainability, plenary at LCM 2017, Luxembourg.
7. Fontes J, Bolhuis A, Bogaers K Saling P, van Gelder R, Traverso M, Das Gupta J, Bosch H, Morris D, Woodyard D, Bell L, van der Merwe R, Laubscher M, Jacobs M, Challis D. Handbook of product social impact assessment. http://product-social-impact-assessment.com. 2016. Accessed 10 Aug 2016.
8. WBCSD, Alvarado C Brown A, Hallberg K, Nieuwenhuizenn P, Saling P, Chan K, Das Gupta J, Morris D, Nicole G, Wientjes F, Dierckx A, Garcia W, Combs C, Kilgore A, Satterfield B, Haver S, Jostmann T, Vornholt G, Bergman U, Feesch J, Whitaker K, Kiyoshi M, Govoni G, Mehta R, Menon A, Sen S, Upadhyayula V, Bande M, Coërs P, Debecker D, Poesch J, Viot JF. Social life cycle metrics for chemical products – A guideline by the chemical sector to assess and report on the social impact of chemical products, based on a life cycle approach, November 2016, www.wbcsd.org/contentwbc/download/1918/24428, Accessed 30 Nov 2017, ISBN 978–2–940521-52-4.
9. United Nations Environment Programme and Society for Environmental Toxicology and Chemistry. Guidelines for social life cycle assessment of products, Paris, 2009.
10. Traverso M, Bell L, Saling P, Fontes J. Towards social life cycle assessment: a quantitative product social impact assessment. Int J Life Cycle Assess. 2016;23:597–606. https://doi.org/10.1007/s11367-016-1168-8.
11. Ecovadis™. http://www.ecovadis.com/de/uber-uns/. Accessed 8 May 2018.
12. Reprisk®. https://www.reprisk.com/. Accessed 8 May 2018.
13. Verisk Maplecroft™. https://www.maplecroft.com/. Accessed 8 May 2018.
14. "Institut für Entwicklung und Frieden (INEF)". Internal working document for human right violation severity. (2015).

Chapter 9
Proposal of Social Indicators to Assess the Social Performance of Waste Management Systems in Developing Countries: A Brazilian Case Study

Valeria Ibañez-Forés, María D. Bovea, and Claudia Coutinho-Nóbrega

Abstract The Brazilian National Solid Waste Policy Law promotes sustainable integrated solid waste management nationally, and is committed to improve "informal" recyclable waste pickers' socio-economic conditions. This has led municipalities to develop waste management strategies to incorporate "informal" waste pickers into the "formal" system. In order to measure the social improvement achieved by this action, it is necessary to define a set of indicators capable of quantifying the social performance of waste management systems that adapt specifically to developing countries.

In this study, a set of social impact categories, indicators and metrics capable of assessing the socio-economic and labour conditions of the different stakeholders involved in the life cycle of a municipal solid waste management (MSWM) system is proposed. Then they are applied to a case study in the city of João Pessoa, Paraíba (Brazil). João Pessoa is one of the pioneering Brazilian cities to incorporate a door-to-door selective waste collection system managed by the previous "informal" waste pickers, reorganised into associations or cooperatives of collectors of recyclable materials. Although this waste collection system has steadily expanded around the city until the present-day, it has never been analysed from a social perspective.

Keywords Waste management · Social life cycle assessment · Developing country · Social indicator

V. Ibañez-Forés (✉) · M. D. Bovea
Department Mechanical Engineering and Construction, Universitat Jaume I, Castellón, Spain
e-mail: vibanez@uji.com

C. Coutinho-Nóbrega
Department of Civil and Environmental Engineering, Universidade Federal da Paraiba, João Pessoa, Brazil

© The Author(s) 2020
M. Traverso et al., *Perspectives on Social LCA*, SpringerBriefs in Environmental Science, https://doi.org/10.1007/978-3-030-01508-4_9

9.1 Introduction

Waste management covers a vast field of human activities which, in developing countries, share some similarities to their social singularities, such as limited participation in selective collection programmes, or waste pickers' poor socio-economic and labour conditions [1].

By taking Brazil as a case study, the National Brazilian Solid Waste Policy Law [2] encourages sustainable integrated solid waste management nationally by improving the working conditions of informal waste pickers by integrating them into formal waste picker cooperatives [3]. Among other actions, this aims to improve the social performance of the Municipal Solid Waste Management (MSWM) systems.

To assess and improve MSWM systems in developing countries, it is necessary to evaluate them from a life cycle perspective, including the assessment of social aspects [4]. However, no consensus has been reached for the social impact assessment method, neither in the impact categories to be used, nor in the stakeholders to be considered [5]. Therefore, in order to analyse the social performance of MSWM systems, it is necessary to define an adequate set of social categories, indicators and metrics and the groups of stakeholders to be taken into account.

With this context in mind, the present research aims to propose and apply a set of social indicators capable of assessing the socio-economic and labour conditions of the different stakeholders involved in the life cycle of an MSWM system in developing countries in general, and in Brazil in particular. To do so, a set of social impact categories, indicators and metrics is proposed after taking into account the needs and characteristics of developing countries and the conclusions drawn from a literature review of social impacts caused by waste management activities worldwide. These proposed social indicators were applied to a case study in João Pessoa (Brazil).

9.2 Methodology

The methodology used for proposing and applying to the case study a set of social indicators and metrics to assess the social performance of MSWM systems in developing countries in general, and in Brazil in particular, consists in the stages showed in Fig. 9.1 and described below.

1. A literature review that focuses on analysing the social impact categories/indicators of system in order to identify those more commonly applied. It has been mostly observed that the reviewed studies are based on the methodological framework proposed by UNEP-SETAC [6, 7].
2. A literature review that focuses on analysing MSWM systems in order to identify the stages, the involved stakeholders, socio-economic and labour conditions, needs, etc. of MSWM systems implemented in developing countries in general, and in Brazil in particular.

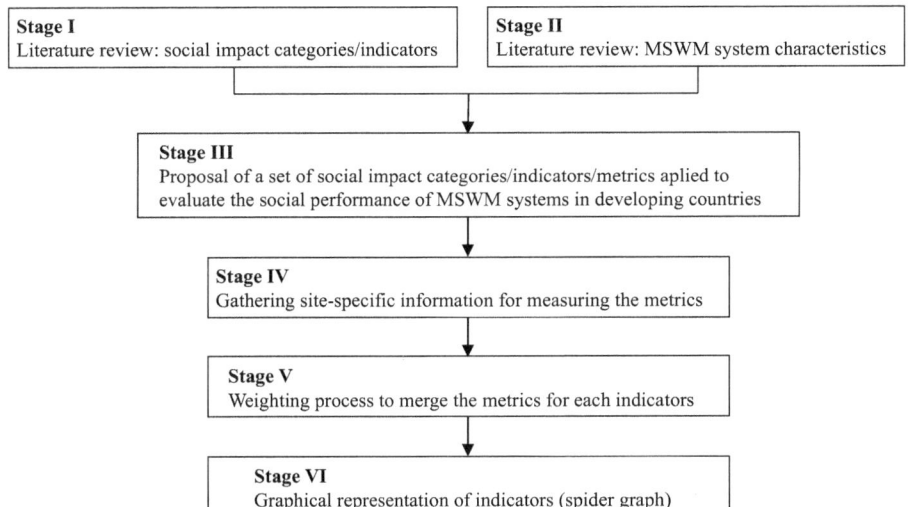

Fig. 9.1 Methodology

3. Proposing a set of social impact categories, with their corresponding indicators, capable of assessing the socio-economic and labour conditions of the stakeholders involved in the life cycle of an MSWM system by taking into account the information from Stages I and II. In order to quantify each indicator, metrics and data source were proposed for each stakeholder that affects it. A set of 12 social impact categories and 22 indicators, with their corresponding metrics, are suggested (for details see Table 9.1).

4. Gathering site-specific information for each metric to apply the proposed social impact categories and indicators to the case study. To do this, questionnaires have to be specifically devised for each group of involved stakeholders.

5. A weighting process need to be applied to merge the metrics for each indicator. A multi-criteria decision analysis is recommended to take into account different preferences for the various social impact categories/indicators [8].

6. The calculated social indicators can be graphically represented to facilitate the identification of the social impact categories/indicators that perform better or worse, or to compare different case studies. To do this, spider graphs may be represented by considering that the bigger the drawn area, the better the system's performance. That is to say, the best social performance of a category would receive the maximum score (100%) and would be plotted on the more external line of the spider graph.

Table 9.1 Proposal of social indicators to assess waste management systems

Social impact category	Social indicators	Metrics	Stakeholder groups			Data source		
			workers	Users	Municipal authorities	Primary data (questionnaire)	Primary data (visits)	Secondary data
1. Working rights	1.1 Freedom for association and collective bargaining	1.1.a Evidence for restrictions to the freedom of association and collective bargaining	•			•		•
		1.1.b Workers have access to meetings and the possibility to dispute resolution procedures	•			•		
		1.1.c Labour union presence	•			•		•
2. Human rights	2.1 Child/senior labour	2.1 Number of children working in the analysed sector	•			•		
3. Quality of job positions (working conditions)	3.1 Fair salary	3.1 Worker salary compared to minimum wage	•			•		
	3.2 Working hours and/ or weekly rest	3.2 Weekly hours actually worked by employees	•			•		
4. Equal opportunities / discrimination	4.1 Gender discrimination	4.1.a Number of women working in waste management	•			•	•	
		4.1.b Gender pay gaps	•			•		
	4.2 Labour regulation	4.2 Number of undocumented workers in waste management	•			•		•
	4.3 Workers from marginal classes	4.3% of workers with no possibility of working in another sector	•			•		
5. Health & safety	5.1 Security & safety for workers	5.1% of workers who use PPEa in their work	•			•		•
	5.2 Long-term health	5.2.a % of vaccinated workers	•			•	•	
		5.2.b % of workers with no health problem	•			•		

Category	Indicator					
6. Working benefits	6.1% of workers with information on the rights that correspond to the waste collector occupational code	•				
	6.2% of workers with the possibility of paying the NHSb	•				•
7. Socio-economic conditions	7.1.a Workers' level of education	•				
	7.1.b Level of education of workers' families	•				
	7.2.a Total monthly family income	•				•
	7.3 Quality of workers/customers' houses	•			•	
8. Community satisfaction & participation	8.1 Social welfare/satisfaction (quality of products/service)	•	•		•	
	8.2% of citizens with access to a reliable WM system	•	•			
9. Value chain actors relationship	9.1.a Customer knowledge about the system	•	•			
	9.1.b Presence of periodical public company reports			•		•
10. Professional development	10.1 Workers' environmental education/awareness	•				

(continued)

Table 9.1 (continued)

Social impact category	Social indicators	Metrics	Stakeholder groups			Data source		
			workers	Users	Municipal authorities	Primary data (questionnaire)	Primary data (visits)	Secondary data
11. Local development (socio-economic repercussion)	11.1 Development of environmental awareness and responsibility	11.1.a Custom environmental awareness		•		•		
		11.1.b % of users receiving environmental information on waste management		•		•		•
	11.2 Local labour integration of formal workers from the informal sector	11.2% of formal workers from the informal sector	•			•		•
12. Governance	12.1 Public commitments to sustainability issues	12.1% of actions made with public funds related to waste management		•	•	•		•
	12.2 Maturity (existence of informal WM system's regulation)	12.2 Legislation on waste management			•			•

[a]PPE – Personal Protective Equipment
[b]NHS - National Health Service

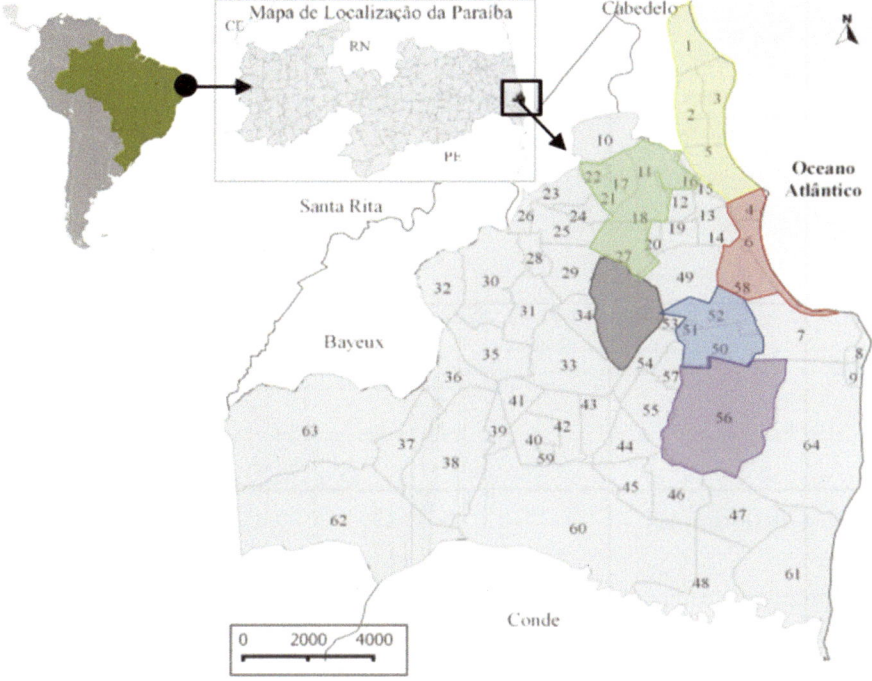

Fig. 9.2 Location of João Pessoa and its districts (Brazil)

9.3   Case Study

The proposed methodology was applied to a case study in the Brazilian city of João Pessoa, whose population (791,000 inhabitants) produces more than 247,000 tons of municipal solid waste per year [9]. The MSWM system in João Pessoa has progressively incorporated a selective collection programme since 2003 [10]. Consequently, different informal waste pickers who previously collected recyclable materials in open dumps have been reorganised in associations that are in charge of door-to-door recyclable material collection and of the manual segregation of recyclable materials in a new manual Material Recovery Facility (MRF) located next to the sanitary landfill. See Fig. 9.2.

To apply the methodology, the site-specific data needed to quantify each metric were collected from the involved stakeholders: workers (waste pickers), users (waste producers) and municipal authorities. To this end, questionnaires were designed for each stakeholder group and were tested by a small sample of surveyed individuals to see if the language was comprehensible, if the response options were suitable and if the necessary information for quantifying the indicators was acquired. Moreover, additional information was obtained from the observations made when visiting the facilities.

The obtained results are presented in Fig. 9.3.

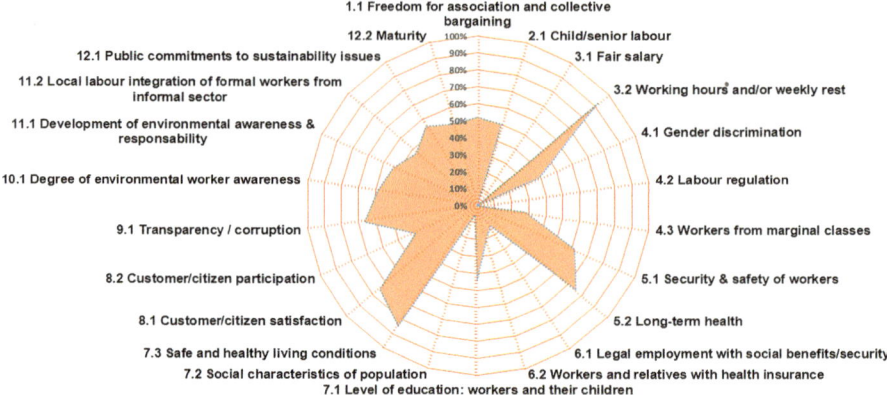

Fig. 9.3 Proposal of the social indicators to assess waste management systems

9.4   Conclusions

The proposed methodology allows the social performance of waste management systems in developing countries to be assessed. This methodology is based on a proposal of social impact categories, and indicators and metrics that facilitate their quantification.

The proposed social impact categories and indicators were applied and validated to the case study of the city of João Pessoa (Brazil). The obtained results demonstrate how the social performance of the MSW management system in João Pessoa still has room for improvement from a social point of view. The better performing social indicators are "Working hours and/or weekly rest" and "Safe and healthy living conditions", followed by "Customer/citizen satisfaction". The worst performing social indicators are "Labour regulation" and "Fair salary", followed by "Social characteristics of population" and "Legal employment with social benefits/security".

Although it has been demonstrated that the proposed social impact categories and indicators are practicable and successful for identifying key aspects and for extending knowledge on the social performance of the MSW management system in the case study, some suggestions for future development are identified. Analysing not only the social characteristics of workers and their families, but those of the users involved in the waste management system, and of the local communities living next to the facilities involved in the system, seems interesting to obtain in-depth knowledge about the effect of the system on its backyard. In line with this, adding an impact category related to some physical impacts, generally considered environmental aspects (e.g. noise, odour, visual impact, etc.), could provide profound knowledge about the analysed systems as these aspects could be important for waste management sites to be accepted by neighbourhoods. Finally, more information about workers' living conditions, such as house size or the security at home, would provide a more detailed picture of the "safe and healthy living conditions" indicator. In addition, previous training about aspects related to occupational diseases, occupational safety and health measures, etc., is highly recommendable to ensure that waste workers correctly understand the questionnaires.

The proposed social impact categories and indicators allow the identification of both the social issues and social benefits associated with the inclusion of the informal waste management sector in municipal management strategies in developing countries, which can help in the decision making that forms part of steering municipal programmes for social performance improvement.

References

1. Aparcana S, Salhofer S. Development of a social impact assessment methodology for recycling systems in low-income countries. Int J Life Cycle Assess. 2013;18:1–10.
2. Lei 12305 de 2 de agosto de 2010. Institui a Política Nacional de Resíduos Sólido, 2010.
3. Magni AAC, Günther WMR. Cooperatives of waste pickers as an alternative to social exclusion and its relationship with the homeless population. Saúde e Sociedade. 2014;23:99–109.
4. Vinyes E, Oliver-Solà J, Ugaya C, Rieradevall J, Gasol CM. Application of LCSA to used cooking oil waste management. Int J Life Cycle Assess. 2013;18:445–55.
5. Yıldız-Geyhan E, Altun-Çiftçioğlu GA, Kadırgan MAN. Social life cycle assessment of different packaging waste collection system. Resour Conserv Recycl. 2017;124:1–12.
6. UNEP-SETAC. Guidelines for social life cycle assessment of products. Paris: United Nations Environment Programme; 2009.
7. UNEP-SETAC. The methodological sheets for subcategories in social life cycle assessment (S-LCA). United nations environment programme; 2013.
8. Azapagic A, Perdan S. An integrated sustainability decision-support framework part II: problem analysis. Int J Sust Dev World. 2005;12:112–31.
9. IBGE. Estimativa da População: Paraíba-João Pessoa. Brazil: Instituto Brazileiro de Geografia e Estatística; 2015.
10. Coutinho-Nóbrega C. Viabilidade Econômica, com Valorização Ambiental e Social, de Sistemas de Coleta Seletiva de Resíduos Sólidos Domiciliares – Estudo de Caso: João Pessoa/PB. Brazil: Universidade Federal de Campina Grande; 2003.

Chapter 10
Social Assessment in the Design Phase of Automotive Component Using the Product Social Impact Assessment Method

Laura Zanchi, Alessandra Zamagni, Silvia Maltese, Rubina Riccomagno, and Massimo Delogu

Abstract This paper shows one of the first example of S-LCA application in the automotive sector by means of the Product Social Impact Assessment method, developed by the Roundtable for the Product Social Metrics. The case study concerns a vehicle component. The main companies involved in the production stage have been engaged in the data collection; therefore, this work gave the opportunity to test the method usability as a supporting tool in the design phase. The main outcomes concern: (i) product system and system boundaries definition; (ii) data collection feasibility; (iii) handbook steps applicability. The PSIA quantitative approach proved to be practicable, even if opportunities for improvements have been identified especially regarding the social indicators granularity in terms of their capability to reflect the differences among the alternative design options from a social point of view. This is a decisive aspect to enhance the assessment of social impacts during the product design phase.

L. Zanchi (✉) · A. Zamagni
Ecoinnovazione srl, Spin-off ENEA, Bologna, Italy
e-mail: l.zamchi@ecoinnovazione.it

S. Maltese
Magneti Marelli S.p.A. – Powertrain Division, Bologna, Italy

Department of Civil, Chemical, Environmental and Materials Engineering, University of Bologna, Bologna, Italy

R. Riccomagno
Magneti Marelli Spa, EHS Central Team, Corbetta, Italy

M. Delogu
Department of Industrial Engineering, University of Florence, Florence, Italy

105

M. Traverso et al., *Perspectives on Social LCA*, SpringerBriefs in Environmental Science, https://doi.org/10.1007/978-3-030-01508-4_10

10.1 Introduction

The automotive sector consists of a complex network of companies which work at different levels of the production stage of a vehicle. It is central to Europe's economic prosperity; however, vehicles are responsible for large-scale environmental and socio-economic impacts at every life cycle stage, which are perceived as the key factor for company's public reputation and attractiveness on the market [1]. The sustainability challenges require the availability of methods and tools able to identify environmental hot spots along the whole value chain and to measure the improvements achieved when new or alternative solutions are designed and implemented. Among the methods, those based on a life cycle approach such as Life Cycle Assessment (LCA), Life Cycle Costing (LCC) and Social LCA (S-LCA), stand out as comprehensive and viable solutions [2–4].

In this context, the experience of Magneti Marelli® during the last 10 years is particularly significant. Within the framework of the Design for Environment principles in the Research and Development (R&D) stage, the Company has been using LCA for 8 years to compare and validate the environmental performances of alternative design options for several components according to the lightweight strategy [5, 6]. The Company product portfolio is represented by components belonging to different systems of the automotive sector (e.g. lighting, powertrain, suspension), so any intervention on them has effects to the whole vehicle too.

Recently, the Company has been involved in facing and discussing also the most challenging aspects of the Life Cycle Thinking methodologies, with the aim of improving methodologies applicability and results usability in the context of sustainable design [7, 8]. For this reason, also Life Cycle Costing (LCC) has been used to compare alternative design solutions with the aim of evaluating the trade-off between production and use stage expenditures when innovative lightweight materials are applied.

Moreover, as the Magneti Marelli Sustainability Program encompasses the social sphere, beyond economic and environmental fields, the Social Life Cycle Assessment (S-LCA) has been also selected by the Company to evaluate social impacts and benefits of its products and to compare alternative design options. In fact, enlarging the product assessment to the three dimensions – environmental, economic and social – increases the awareness of the company's impacts within society and support a transparent and reliable communication toward stakeholder.

In the recent years the Roundtable for Product Social Metric initiative has developed a handbook which proposes a practical method named Product Social Impact Assessment (PSIA) for the quantification of the social performances [9, 10]. Starting out from the Social Life Cycle Assessment guidelines [11,12], the PSIA proposes a defined list of indicators and qualitative and quantitative methods for their quantification, with the ultimate goal to practically support organizations into assessing the potential social impacts of a product along its life cycle [10].

In this work, the PSIA method is applied to assess the social performance of a vehicle component, produced by Magneti Marelli. The main reasons behind the selection of such method are the following: the interest on evaluating a social assessment at product level, so as to complement the LCA and LCC evaluations already developed by the Company; the possibility to start such analysis from a defined and manageable list of indicators; the possibility of elaborating data by means of a clear process.

The main objective of this work is to test the PSIA applicability and the outcomes usability when the social assessment is carried out at an early design phase to identify distinctive feature of alternative design solutions.

10.2   Materials and Method

10.2.1   Case Study Description

The case study concerns a part of the suspension system of a vehicle named knuckle (a steel made part with a total weight of around 6 kg), produced by Magneti Marelli. The Magneti Marelli's plant produces different customised knuckles for different car makers; in this study a specific model has been selected in order to focus on a specific supply chain. The production stage, from raw materials supply to assembly to the vehicle, involves several processes carried out by four companies and four different plants.

The component is produced starting from steel scraps, which are mainly provided by companies in Poland, while virgin steel represents only a minor part. Raw steel is firstly worked by company (A)[1] which is in charge of the casting process; then a second company (B) developed the surface treatment and cathaphoresis process of the semi-worked piece. The component goes to the plant of Magneti Marelli (company C) for the final machining and quality control before it is sent to the car maker (company D) plant, located in Germany, where it is assembled to the suspension system and the vehicle. These first three production steps are developed by three different companies and plants within the same industrial district located in Bielsko-Biala (Poland). For this reason, the transports among the different plants are negligible.

Overall, the bill of material of the knuckle consists of a few key materials, which are manufactured by means of mature technologies; nevertheless, the quite fragmented supply chain (in particular the steel scrap supply chain) could represent a critical aspect in terms of companies' involvement in the data collection and relationships management among the different stakeholders along the whole life cycle. Therefore, in this study the four companies (A, B, C and D) were actively involved for the collection of primary data at their sites for the performance

---

[1]For confidentiality reasons, the name of the companies is not displayed.

indicators related to workers and local communities. Afterwards, they were involved also in providing feedbacks about the proposed indicators selected for the analysis. The knuckle life cycle stages, and related companies, and the system boundaries are represented in Fig. 10.2.

10.2.2 Product Social Impact Assessment

The Product Social Impact Assessment method (PSIA) concerns social topics and performance indicators that reflect both positive and negative performance of the product on three stakeholder groups: workers, consumers and local communities. Two approaches are proposed: quantitative and qualitative [10]. Within an overall framework based on the typical four LCA phases, the PSIA method (quantitative approach) compels the steps depicted in Fig. 10.1.

The Goal and Scope definition step concerns the definition of: (1) the goal (i.e. steer product development) and the product being assessed clarification; (2) the geographic scope of the assessment, in terms of value chain actors included in the assessment, and their respective sectors and locations; (3) stakeholder groups on which the assessment is focused. Regarding the subsequent inventory step, the handbook introduces two approaches for the assessment; the first, named quantitative, which collects only numerical data, whereas the second, named scales-based, which takes into consideration both quantitative and qualitative data. In this study the quantitative approach has been selected: it uses only numerical data measured as performance indicators, grouped into several social topics referred to three stakeholder groups (Table 10.1). Each social topic is represented by one or two performance indicators in the form of cost (the lowest is the positive) or benefit (the highest is the positive). The quantitative indicators are measured as absolute numbers (e.g. number of actions) or percentages (e.g. % of workers) (Table 10.1). Indicators listed in (see Table 10.1) were collected for each company/life cycle stage (LCSi), for an overall amount of 19 social topics measured by means of 30 performance indicators.

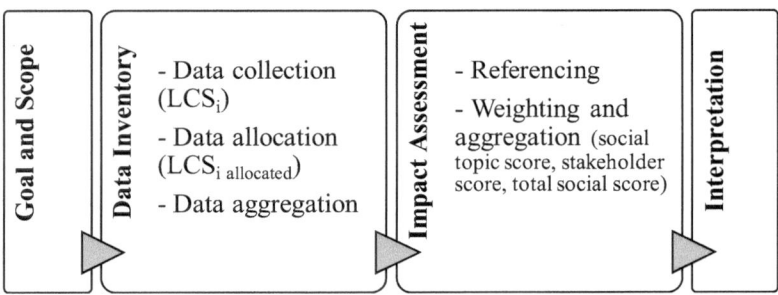

Fig. 10.1 Steps of the Product Social Impact Assessment (PSIA) method, quantitative approach

Table 10.1 Stakeholder groups, social issues and performance indicators of the quantitative approach [9] (Benefit (B) = higher is better; Cost (C) = lower is better)

	Social topics	Performance indicators	Unit	Type
Workers	Health and safety	Number of hours of health & safety training given during the reporting period.	Hours	Benefit
		Average number of incidents during the reporting period.	Number	Cost
	Wages	Percentage of workers whose wages meet at least the legal or industry minimum wage and their provision fully complies with all applicable laws.	%	Benefit
		Percentage of workers who are paid a living wage.	%	Benefit
	Social benefits	Percentage of workers whose social benefits meet at least legal or industry minimum standards and their provision fully complies with all applicable laws.	%	Benefit
	Working hours	Percentage of workers who exceeded 48 h of work per week regularly during the reporting period.	%	Cost
	Child labour	Number of hours of child labour identified during the reporting period.	Hours	Benefit
		Number of actions during the reporting period targeting business partners to raise awareness of the issue of child labour.	Actions	Cost
	Forced labour	Number of hours of forced labour identified during the reporting period.	Hours	Benefit
		Number of actions during the reporting period targeting business partners to raise awareness of the issue of forced labour.	Actions	Cost
	Discrimination	Number of complaints identified during the reporting period related with discrimination.	Complaints	Benefit
		Number of actions taken during the reporting period to increase staff diversity and/or promote equal opportunities.	Actions	Cost
	Freedom of association and collective bargaining	Percentage of workers identified during the reporting period who are members of associations able to organise themselves and/or bargain collectively.	%	Benefit

(continued)

Table 10.1 (continued)

	Social topics	Performance indicators	Unit	Type
	Employment relationship	Percentage of workers who have documented employment conditions.	%	Benefit
	Training and education	Number of hours of training per employee during the reporting period.	Hours	Benefit
	Work-life balance	Percentage of workers with direct family responsibilities who were eligible for maternity protection. Or to take maternity. Parental. Or compassionate leave during the reporting period.	%	Benefit
	Job satisfaction and engagement	Percentage of workers who partici-pated in a job satisfaction and engagement survey during the reporting period.	%	Benefit
		Worker turnover rate during the reporting period.	%	Cost
Consumers	Health and safety	Number of claims acknowledged by a certification or accreditation body that the product contributes to a higher level of consumer health or safety.	Claims	Benefit
		Number of complaints identified during the reporting period related to consumer health and safety.	Complaints	Cost
	Experienced Well-being	Composite measure of experienced Well-being (1–10)	Absolute metric	Benefit
Local communities	Health and safety	Number of programmes during the reporting period to enhance com-munity health and safety.	Programmes	Benefit
		Number of adverse impacts on community health or safety identi-fied during the reporting period.	Adverse impacts	Cost
	Access to tangi-ble resources	Number of programmes during the reporting period to enhance com-munity access to tangible resources or infrastructure.	Programmes	Benefit
		Number of adverse impacts on community access to tangible resources or infrastructure during the reporting period.	Adverse impacts	Cost
	Local capacity building	Number of programs targeting capacity building in the community during the reporting period.	Programmes	Benefit
		Number of people in the commu-nity benefitting from capacity	Persons	Benefit

(continued)

Table 10.1 (continued)

	Social topics	Performance indicators	Unit	Type
		building programmes during the reporting period.		
	Community engagement	Number of programmes or events targeting community engagement during the reporting period.	Programmes	Benefit
	Employment	Number of new jobs created during the reporting period.	New jobs	Benefit
		Number of jobs lost during the reporting period.	Jobs lost	Cost

When a quantitative approach is applied, data allocation to the product is a fundamental step and the handbook suggests to apply an allocation factor based on the number of working hours needed to produce one unit of the product (Eq. 10.1).

$$allocation\,factor = \frac{P_{Site} \times h_{empl.site} \times 52weeks}{P_{production\,line} \times h_{empl.production\,line} \times 52weeks} \qquad (10.1)$$

Where:

- P site is the number of employees at the site;
- h empl. Site is the average number of working hours per employee per week at the site;
- P production line is the number of employees working at the specific production line;
- h empl. Production line is the average number of working hours per employee per week at the production line

Once performance indicators values of each life cycle stage are allocated to the product (more precisely the output from the given stage), then the allocated values are aggregated along the life cycle stages (PLC indicator). The handbook provides two formulas for the aggregation according to the nature of indicator (absolute number or percentage). When inventory data are aggregated, the impact assessment is carried out; this phase consists in the social assessment method type I [13] proposed by the Roundtable, which elaborates the aggregated values according to a referencing step. The referencing consists of calculating a performance value (positive or negative) for each indicator by comparing the PLC indicator with a corresponding reference value. The handbook suggests three different referencing processes, depending on the indicator nature, and reference values for each performance indicator according to an ideal/worst scenario. The last step of the PSIA method is the weighting and aggregation of the product social performances to the different social topics first and, afterwards, to stakeholder groups up to a total social score. In the quantitative approach, aggregation of performance indicators into social topic scores, stakeholder scores and total social score is only possible when comparing alternative products. In this study only one product is assessed therefore no comparison is involved and the aggregation of performance indicators is not carried out.

10.3 Results

10.3.1 Goal and Scope Definition

The main elements mentioned in the handbook concerning goal and scope phase are the geographic scope and the stakeholder groups. However, to practically carry out the data collection at site level and the following data elaboration and interpretation, a clear definition of the product system and system boundaries is needed. In this regard, the approach described on Zanchi et al. [2] has been applied; which consists of defining the product system according to a technology-oriented (identification of the several separated technological units positioned throughout the product life cycle) and an organization-oriented (identification of the several companies positioned throughout the product life cycle) approach. As for the system boundaries definition, the double-layer approach has been used [2]; it consists of a physic layer (identifying the processes included in the analysis) and an effect layer (providing the stakeholder groups considered along the product life cycle). In this work the system boundaries mainly include processes/companies involved in the production stage and two stakeholder groups (workers and local communities) related to these stages (see Fig. 10.2).

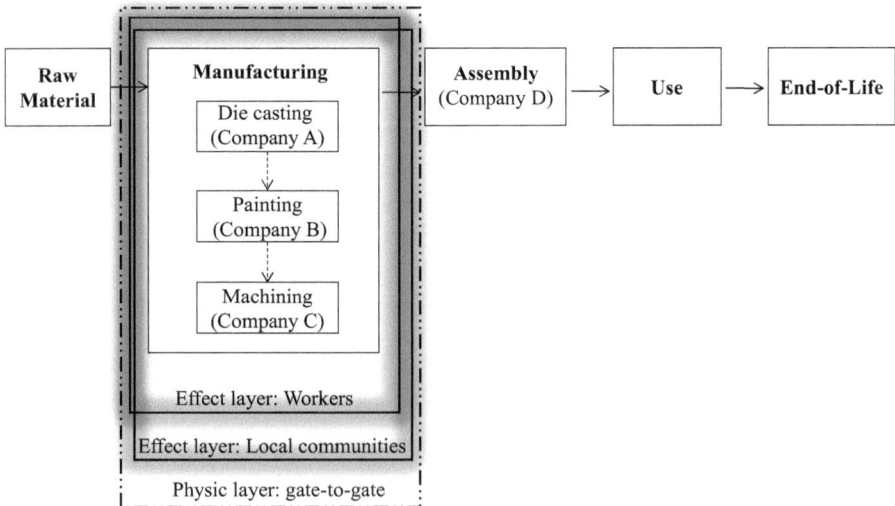

Fig. 10.2 Life cycle stages and system boundaries of S-LCA of knuckle

10.3.2   Data Inventory: Relevance, Affordability and Completeness

Companies were invited to collect primary data at their sites and, afterwards, to discuss and provide feedback about the proposed indicators. The general feedback received could be summed up in three main groups: relevance, affordability and completeness. In terms of relevance two aspects came out: the stakeholder "consumer" is considered misleading since it is not clear the target for the given product; the topic "Health and safety" for local communities is considered not relevant for the specific activities carried out by the specific plants. In terms of affordability three considerations were collected. The first regards the privacy that could, in some case, hamper data collection (e.g. "Freedom of association and collective bargaining"). Secondly, measures about some topics (e.g. "child labour" and "discrimination") could mainly regard material suppliers or those companies whose activities are developed in areas/countries where such social aspects are not ruled by specific laws. In this case, existing declarations and claims provided by suppliers were not directly managed by persons responsible of plant, involved in the data collection, so it was found difficult to give evidence of them. All the indicators regarding actions/programmes were found difficult to be interpreted. For example, in some case it was not easy to distinguish actions for "Local capacity building" from programmes regarding "Community engagement". In this regard the companies were asked to provide examples of actions/programmers that they included in the questionnaire in order to provide more insights. Besides some difficulties regarding the understanding of some specific indicators, other comments regarded the completeness of the list which was found lacking of some social aspects that companies are called to manage.

10.3.3   Referencing

According to the PSIA, the impact assessment phase consists of the comparison of the allocated social indicator values with reference values, followed by a weighting procedure to obtain social topics score, stakeholders score and total social score [10]. In particular, the allocated values are first aggregated along the whole product life cycle, then they are compared to reference values in order to evaluate the performance value (positive or negative). Identifying proper reference values is an important aspect but in this case it has been excluded from the scope of this work. Therefore, in this study the reference value proposed by the Handbook have been applied as a way to test practicability of the whole procedure and to find out the nature and usability of the outcomes by the Magneti Marelli. Table 10.2 reports referencing procedure outcomes which are represented by a qualitative performance evaluation of each indicator.

Table 10.2 S-LCA impact assessment results: performance evaluation of knuckle

Performance indicators	Unit	Performance evaluation
Number of hours of health & safety training given during the reporting period.	Hours	Negative performance
Average number of incidents during the reporting period.	Number	Negative performance
Percentage of workers whose wages meet at least the legal or industry minimum wage and their provision fully complies with all applicable laws.	%	Target or minimum scenario has been reached
Percentage of workers who are paid a living wage.	%	Target or minimum scenario has been reached
Percentage of workers whose social benefits meet at least legal or industry minimum standards and their provision fully complies with all applicable laws.	%	Target or minimum scenario has been reached
Percentage of workers who exceeded 48 h of work per week regularly during the reporting period.	%	Target or minimum scenario has been reached
Number of hours of child labour identified during the reporting period.	Hours	Target or minimum scenario has been reached
Number of actions during the reporting period targeting business partners to raise awareness of the issue of child labour.	Actions	Negative performance
Number of hours of forced labour identified during the reporting period.	Hours	Target or minimum scenario has been reached
Number of actions during the reporting period targeting business partners to raise awareness of the issue of forced labour.	Actions	Negative performance
Number of complaints identified during the reporting period related with discrimination.	Complaints	Target or minimum scenario has been reached
Number of actions taken during the reporting period to increase staff diversity and/or promote equal opportunities.	Actions	Negative performance
Percentage of workers identified during the reporting period who are members of associations able to organise themselves and/or bargain collectively.	%	Positive performance
Percentage of workers who have documented employment conditions.	%	Target or minimum scenario has been reached
Number of hours of training per employee during the reporting period.	Hours	Negative performance
Percentage of workers with direct family responsibilities who were eligible for maternity protection. Or to take maternity. Parental. Or compassionate leave during the reporting period.	%	Positive performance
Percentage of workers who participated in a job satisfaction and engagement survey during the reporting period.	%	Positive performance
Worker turnover rate during the reporting period.	%	Negative performance
Number of programs targeting capacity building in the community during the reporting period.	Programmes	Negative performance

(continued)

Table 10.2 (continued)

Performance indicators	Unit	Performance evaluation
Number of people in the community benefitting from capacity building programmes during the reporting period.	Persons	Negative performance
Number of programmes or events targeting community engagement during the reporting period.	Programmes	Negative performance
Number of new jobs created during the reporting period.	New jobs	Negative performance
Number of jobs lost during the reporting period.	Jobs lost	Target or minimum scenario has been reached

10.4 Discussion

The application of the PSIA method to a real case study pointed out some key aspects related to both the applicability of the method and relevance of the results achieved. The first concerns the indicators capability to provide social results sufficiently detailed when the method is applied to a vehicle component; LCA and LCC results generally provide consistent results at component level and enable comparison of design alternatives because are able to reflect technical differences related to materials or manufacturing technologies. Outcomes from this study suggest that this is not guaranteed by the social indicators, which in some case could not be able to reflect those differences in terms of social sustainability. The quantitative nature of the method has the advantage to provide results that could be integrated with the environmental and economic ones, carried out following the same approach; however, in some cases the quantitative nature of some indicators risks hindering differences among alternatives. For example, the number of programmes targeting community engagement does not provide full information about the extent of this action. In this regard, other approaches seem to be more appropriate to identify this aspect (e.g. Sustainable Return on Investment).

Another point of discussion is the number and appropriateness of the indicators. Feedback from companies suggest that the list proposed by the Handbook is a good starting point but a review is needed to make the method fully applicable since the design phase. Moreover, this revision should take into account the already existing CSR strategies and the Key Performance Indicators identified by the company. The Social LCA could then provide a structure approach, within which also the CSR elements and features are framed and evaluated, thus increasing the consistency of the approaches for dealing and measuring the social performances at product and organisational level.

10.5   Conclusion and Outlook

The objective of the study was to test the S-LCA capability, according to the PSIA method, to support the design process, towards a sustainable design approach. The application suggests that key elements that would favour this applicability are the quantitative and semi-quantitative nature of the PSIA method, that could also enhance its combination with other life cycle-based methodologies (e.g. LCA and LCC). The applied PSIA approach proved to be practicable, even if opportunities for improvements have been identified. The first concerns how to properly set the boundaries of the system analysed (i.e., how far in the value chain should we go when accounting for the social impacts). Secondly, the choice of social indicators, on the basis of their relevance for the organisational system at hand and their capability to reflect the differences among the alternative design options from a social point of view. A proposal for setting the system boundaries has been introduced, while for indicators, it is recognised the need for setting up more structured and participative approach with a representative set of stakeholders. An increased guidance has been envisaged as necessary on this aspect, as it would favour also an increased and active involvement of the organisations themselves.

References

1. Koplin J, Seuring S, Mesterharm M. Incorporating sustainability into supply management in the automotive industry – the case of the Volkswagen AG. J Clean Prod. 2007;15:1053–62. https://doi.org/10.1016/j.jclepro.2006.05.024.
2. Zanchi L, Delogu M, Zamagni A, Pierini M. Analysis of the main elements affecting social LCA applications: challenges for the automotive sector. Int J Life Cycle Assess. 2018;23 (3):519–35. https://doi.org/10.1007/s11367-016-1176-8.
3. Traverso M, Bell L, Saling P, Fontes J. Towards social life cycle assessment: a quantitative product social impact assessment. Int J Life Cycle Assess. 2018;23(3):597–606. https://doi.org/10.1007/s11367-016-1168-8.
4. Tarne P, Traverso M, Finkbeiner M. Review of life cycle sustainability assessment and potential for its adoption at an automotive company. Sustainability. 2017;9:670. https://doi.org/10.3390/su9040670.
5. Delogu M, Maltese S, Del Pero F, Zanchi L, Pierini M, Bonoli A. Challenges for modelling and integrating environmental performances in concept design: the case of an automotive component lightweighting. Int J Sustain Eng. 2018;11:135–48. https://doi.org/10.1080/19397038.2017.1420110.
6. Maltese S, Delogu M, Zanchi L, Bonoli A. Application of Design for Environment principles combined with LCA methodology on automotive product process development: the case study of a crossmember. Sustainable Design and Manufacturing 2017, Smart Innovation, Systems and Technologies 68, 2017.
7. Delogu M, Zanchi L, Maltese S, Bonoli A, Pierini M. Environmental and economic life cycle assessment of a lightweight solution for an automotive component: a comparison between talc-filled and hollow glass microspheres-reinforced polymer composites. J Clean Prod. 2016;139:548–60. https://doi.org/10.1016/j.jclepro.2016.08.079.

8. Delogu M, Zanchi L, Dattilo CA, Maltese S, Riccomagno R, Pierini M. Take-home messages from the applications of life cycle assessment on lightweight automotive components, SAE international, CO_2 reduction for transportation systems conference Turin, Italy, 6–7-8 June 2018.
9. Fontes J, Bolhuis A, Bogaers K, Saling P, van Gelder R, Traverso M, Tarne P, Das Gupta J, Morris D, Woodyard D, Bell L, van der Merwe R, Kimm N, Santamaria C, Laubscher M, Jacobs M, Challis D, Alvarado C, Duclaux C, Slaoui Y, Culley H, Zinck S, Stermann R, Carteron E, Gupta A, Nilsson S, Gaasbeek A, Goedkoop M, Evitts S. Handbook for product social impact assessment version 3.0. 2016. http://product-social-impact-assessment.com/wp-content/uploads/2014/08/Handbook-for-Product-Social-Impact-Assessment.pdf.   (Accessed 16.05.2018).
10. Fontes J, Tarne P, Traverso M, Bernstein P. Product social impact assessment. Int J Life Cycle Assess. 2018;23(3):547–55. https://doi.org/10.1007/s11367-016-1125-6.
11. United Nations Environment Programme and Society for Environmental Toxicology and Chemistry, Guidelines for social life cycle assessment of products, Paris, 2009.
12. United Nations Environment Programme and Society for Environmental Toxicology and Chemistry, The methodological sheets for sub-categories in social life cycle assessment (S-LCA). UNEP-SETAC life-cycle initiative, Paris, France, 2013.
13. Garrido SR, Parent J, Beaulieu L, Revéret J-P. A literature review of type I S-LCA—making the logic underlying methodological choices explicit. Int J Life Cycle Assess. 2018;23(3):432–44. https://doi.org/10.1007/s11367-016-1067-z.

Chapter 11
Social Life Cycle Assessment in Agricultural Systems – U.S. Corn Production as a Case Study

Markus Frank, Thomas Laginess, and Jan Schöneboom

Abstract Socio-Economic Life Cycle Assessment (S-LCA) has proved to be a useful approach for quantitative sustainability assessment. A sustainability assessment method developed by BASF, AgBalance™, includes primary agricultural production that integrates environmental life cycle assessment (LCA), S-LCA and economic cost considerations with quantitative sustainability indicators. It is based on mandatory and optional parts of the ISO 14040 and 14,044 standards (2006) for life cycle assessment. Furthermore, the guidelines of the UNEP/SETAC working group for S-LCA as well as the SA8000 and ISO26000SR standards were followed in the development of the methodology. In a case study, a decade of corn production in Iowa was analyzed in order to compare the sustainability of agricultural practices (Year 2000 vs. Year 2010). The integrated impacts of social indexes in the Iowa farming community yielded a substantial increase in the sustainability performance, mainly driven by the indicators Professional Training, Succession and Gender Equality. In summary, this case study underlines the paradigm of sustainable intensification.

11.1 Introduction

Social impacts are not addressed specifically in the ISO LCA standards, and there are no other consensus standards that can be referenced to define the criteria for a S-LCA. AgBalance™ represents an approach to create a S-LCA framework through the identification and use of relevant factors associated with life cycle principles. Even though there are no industry standards available, the guidelines from the

M. Frank (✉)
BASF SE – Crop Protection, Sustainability Assessment, Agricultural Center, Limburgerhof, Germany
e-mail: markus.frank@basf.com

T. Laginess
BASF Corporation, Applied Sustainability, Wyandotte, MI, USA

J. Schöneboom
BASE SE, Applied Sustainability, Ludwigshafen, Germany

© The Author(s) 2020
M. Traverso et al., *Perspectives on Social LCA*, SpringerBriefs in Environmental Science, https://doi.org/10.1007/978-3-030-01508-4_11

UNEP/SETAC working group [1] as a starting point. The social assessment in AgBalance™ is based on the SEEBALANCE® scheme for S-LCA, which was developed in 2005 by the Universities of Karlsruhe and Jena, the Öko-Institut Freiburg e.V., and BASF respectively [2, 3]. This approach to social assessment is based on a sectoral approach where key social figures from different industry segments are related to their corresponding production volumes. The resulting social profiles for processes or products then assume a format, equivalent to the LCI in the environmental section. For all social indicators, the production volumes are related quantitatively to a given industry sector (e.g., 'occupational diseases per kg product'). With this approach, it is possible to relate the inputs and outputs from the environmental life cycle assessment to the individual social indicators. To this end, different statistical databases are combined to connect social indicators to production volumes. The link between products and corresponding social impacts is made by a sector assessment. This is based on either the 'Nomenclature Générale des activités économiques dans les Communautés Européennes' (NACE, general nomenclature of economic activities in the European Community) – an initiative that classifies all industries into different sectors – or the ISIC, the International Standard Industrial Classification. All products can be linked to these NACE/ISIC codes, using the product classification list (CPA = Classification of Products by Activity). Using statistical data for both production volumes and e.g. working accidents, a database for each industry sector can be created. This procedure is repeated for every AgBalance™ indicator. When comparisons between national currencies are made, all monetary quantities are adjusted, using purchasing power parity. In an AgBalance™ study, the social impacts are quantified, according to the functional unit, and aggregated for all up- and downstream life cycle stages [4].

During the development process, concrete targets for social sustainability for products and processes were derived. This was done through analysis of more than 60 published studies on the social topics by various institutions. As a result, more than 700 goals and more than 3200 indicators were systematically recorded, categorized and summarized. For AgBalance™, this set of social parameters has been extended and in parts modified, to address specific agricultural sustainability topics, e.g., access to land, the level of organization or international trade with agricultural products. These topics were initially identified through a stakeholder process in 2009 and 2010, organized by BASF, and were subsequently discussed with leading experts. Feedback from this process was then integrated into the development of these indicators.

Social impacts are aggregated, based on normalization, relevance and societal weighting factors to form the following stakeholder impact categories and indicators:

11.1.1 Stakeholder Category: Employee/Farmer

(1) Working accidents and fatal working accidents (number per CB)
Negative indicator – lower numbers are seen to be better.
The number of working accidents is recorded in association with an activity (production).

(2) Occupational diseases (number per CB)
Negative indicator – lower numbers are seen to be better.
The number of occupational disease is recorded in association with an activity (production).

(3) Human toxicity (toxicity score per CB)
Negative indicator – lower numbers are seen to be better.
The assessment of life cycle toxicity potential is based on the framework for the toxicity potential assessment is described in [5].

(4) Wages and salaries (monetary value per CB)
Positive indicator – higher numbers are seen to be better.
This indicator evaluates the wages for people in (industrial) upstream and downstream processes.

(5) Professional Training (monetary value per CB)
Positive indicator – higher numbers are seen to be better.
This indicator evaluates professional training, i.e. informal education in the respective industry sectors for upstream and downstream.

(6) Strikes and lockouts (lost working hours per CB)
Negative indicator – lower numbers are seen to be better.
Freedom to assemble and a guarantee of human rights are assumed to be preconditions that must be fulfilled.

11.1.2 Stakeholder Category: Consumer

(1) Residues in feed and food (performance rating, percentage maximum residue level exceedance)
Negative indicator – lower numbers are seen to be better.
The indicator assesses the percentage of food samples that exceed official maximum residue limits (MRLs).

(2) Presence of unauthorized/unlabeled GMO in feed and food (performance rating, number of occurrences)
Negative indicator – lower numbers are seen to be better.
This indicator is based on a retrospective analysis of reported occurrences of unlabeled or unauthorized residues of genetically modified organisms (GMO) in food products (based on official food monitoring reports).

11.1.3 Stakeholder Category Local and National Community

(1) Employment (working years per CB)
Positive indicator – higher numbers are seen to be better.
This indicator evaluates the contribution that the product system makes to employment and job creation.

(2) Qualified employees (working years per CB)
Positive indicator – higher numbers are seen to be better.
This indicator calculates the working time that qualified employees with a formal degree dedicate to a specific product system versus unskilled worker.

(3) Gender equality (working years per CB)
Positive indicator – higher numbers are seen to be better.
In the assessment of upstream and downstream industrial production steps, this indicator is calculated by considering the number of female managers (higher level) in the respective industry sectors.

(4) Integration of disabled employees (working years per CB)
Positive indicator – higher numbers are seen to be better.
This indicator assesses the employment rate for people with severe disabilities in upstream and downstream processes that are part of the product system.

(5) Access to land (monetary value per CB)
Negative indicator – lower numbers are seen to be better.
This indicator therefore calculates the percentage of leased land – within the agricultural area – that is used for the benefit of the customer, multiplied by the cost of the lease.

(6) Family support (monetary value per CB)
Positive indicator – higher numbers are seen to be better.
This indicator evaluates – in financial terms – the impact of parental leave and other bonuses offered to employees, who are married and/or have children, including health insurance and support for births, deaths etc.

11.1.4 Stakeholder Category: International Community

(1) Imports from developing countries (monetary value per CB)
Positive indicator – higher numbers are seen to be better.
This indicator rates the monetary value associated with the import of raw-materials, industrial goods etc., that are part of the product system for upstream and downstream processes. As it contributes to the income of local producers, it supports the economy in the developing region.

(2) Fair trade benefits (monetary value per CB)
Positive indicator – higher numbers are seen to be better.
This indicator calculates the summary of benefits, such as guaranteed prices and premiums, paid to producers for each alternative that is associated with the same customer benefit.

11.1.5   Stakeholder Category: Future Generations

(1) Number of trainees (number of persons per CB)
Positive indicator – higher numbers are seen to be better.
This indicator assesses the number of people in formal education within the industrial sectors, associated with the relevant upstream and downstream processes.

(2) R&D expenditures (monetary value per CB)
Positive indicator – higher numbers are seen to be better.
This indicator quantifies the internal and external expenditure of companies in R&D activities.

(3) Capital investment (monetary value per CB)
Positive indicator – higher numbers are seen to be better.
This definition covers the value of replacement and net investment, including general repair, purchase of concessions, patents and licenses.

(4) Social Security (monetary value per CB)
Positive indicator – higher numbers are seen to be better.
This assessment summarizes the payments employers make to health insurance schemes and unemployment insurance, pensions and similar programs for their employees.

In the AgBalance™ methodology, environmental, social and economic impacts are first assessed independently. The social impact assessment uses characterization factors (as in most LCIA methods) with the resulting impacts normalized to arrive at the individual impact categories. The normalized results for different impact categories are represented as the fingerprint for each alternative. Relative improvement in each impact is represented by smaller values on the respective axes; hence the smaller the fingerprint, the better the relative performance of the corresponding alternative (Fig. 11.1 and Fig. 11.2). Using relevance and societal weighting factors, they are then further combined into a single social score impact as shown in Fig. 11.2 (illustrated as the so-called socio-eco-efficiency score; [4]). The relevance factor reflects the extent to which a given environmental or social impact, e.g., emission, energy consumption or working accidents, contributes to the total burden in a given geographic region. Where appropriate, relevance factors are also calculated for social metrics. The relevance factors are updated at least every 7 years or more frequently, as deemed necessary.

The weighting factors for used in this case study are summarized in Table 11.1.

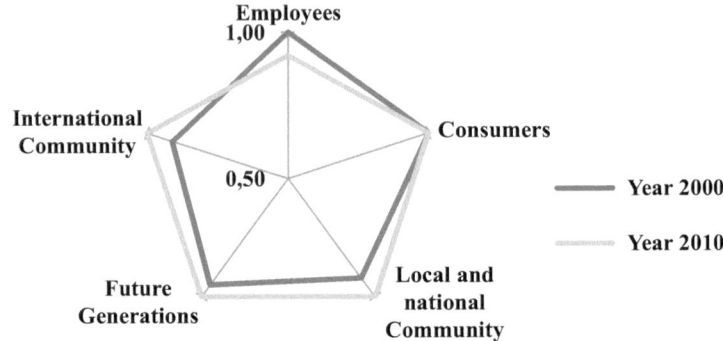

Fig. 11.1 Social Fingerprint of prechain and agricultural production of corn production in Iowa 2000 & 2010. Smaller figures represent a lower impact

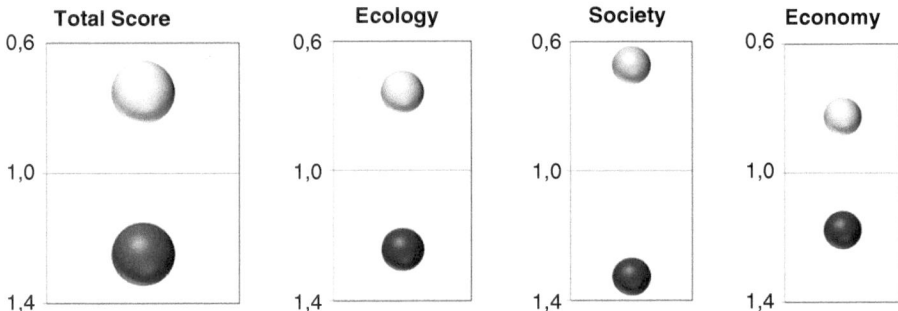

Fig. 11.2 Total socio-eco-efficiency score of Iowa corn production in 2000 (dark gray) and 2010 (light gray) in the AgBalance™ assessment. Smaller figures represent a lower impact

11.1.6   Case Study – Corn production in Iowa

In a case study, a decade of corn production in Iowa was analyzed in order to compare the sustainability performance of contemporary farming with former agricultural practices (2000 vs. 2010). It has been reported that the economics of corn production have largely improved over the last decade [6]. The introduction of improved corn hybrids with better agronomics and biotech traits, the replacement of organophosphate soil insecticides through new-generation seed treatments and more efficient use of larger fertilizer input rates has resulted in a substantial increase in the average yield per hectare. It was unclear, however, whether these improvements came at the expense of the sustainability performance of the current agricultural practice.

The goal defined for the AgBalance™ case study was to quantify the differences in life cycle environmental impacts, total life cycle costs and social aspects of corn production systems in the United States. The Customer Benefit (CB; functional unit)

Table 11.1 Social weighting factors used in the AgBalance™ case study

Employee/Farmer 25%	Working accidents 15%
	Fatal working accidents 20%
	Occupational Diseases 15%
	Toxity Potential 25%
	Wages 10%
	Professional Training 10%
	Organization 5%
Consumer 20%	Residues in feed and food 60%
	Residues of GMO 40%
Local/National Community 25%	Access to land 50%
	Employment 20%
	Gender equality 20%
	Integration 10%
International community 10%	Imports from Devel. Countries 66%
	Fair trade 33%
Future generations 20%	Trainees 50%
	Social security 50%

applied to all alternatives for the base case analysis is the evaluation of the inputs required to produce one metric ton (1000 kg) of corn in the state of Iowa, which is equivalent to 39.4 bushels of corn (56 lb. per bushel of corn) in one growing season (1 year). The two alternatives chosen were the average agricultural practice in Iowa in the year 2000 compared to 2010. Most of the data used in the study were derived from Iowa State University research on corn production [7, 8]. The environmental impacts for the production of the two alternatives were calculated from life cycle inventories for the input parameters such as fuel usage, fertilizers and pesticides. Life cycle inventory data were from several data sources, such as ecoinvent 2.0, Boustead database and BASF's manufacturing database. Overall, the quality of the data was considered medium-high to high. None of the life cycle inventory data was considered to be of low data quality [9].

The major factor influencing the environmental and cost impact between the 2 years is the yield increase in the production of corn. Iowa State University data shows an increase of 21.7% from 2000 to 2010 in corn production yield [7, 8]. This information by the University's extension service was based on average data collected for the specific years. Downstream processes such as transportation, drying, storage, processing and secondary uses were excluded from the study as they can be considered equal for both alternatives. The justification for these boundaries is that these are the major impact categories for the production of corn and the only difference between the two alternatives is the data used for the different years. The use and disposal of the corn was not evaluated because the CB of one metric ton for both alternatives was the same. The eco-toxicity potential of the input chemicals is defined to be evaluated in the use phase of the input chemicals only (i.e. the agricultural production).

11.2 Social Fingerprint of Corn Production in Iowa

The assessment of social impacts in the up- and downstream processes in AgBalance™ is based on the SEEBALANCE® method [2]. This approach to social assessment uses a sectoral assessment, where key social figures from different industry segments are related to their corresponding production volumes. The resulting social profiles for processes or products then assume a format, equivalent to the eco-profiles, used in the environmental part. Table 11.2 summarizes the social fingerprint data and Fig. 11.2 for the corn production social fingerprint.

For all social indicators, a quantitative relationship is made for the production volumes of a given industry sector (e.g. "occupational diseases per kg product"). With this approach, it is possible to relate the inputs and outputs from the environmental life cycle assessment to the social indicators. To this end, different statistical databases are combined to connect social indicators to production volumes. The link between products and corresponding social impacts is made by a sector assessment. It is based on the 'Nomenclature générale des activités économiques dans les

Table 11.2 Social fingerprint values for corn production in Iowa 2000 & 2010

Impact category	Indicator	Unit	Year 2000	Year 2010
Employee/farmer	Working hours	h / CB	0.713	0.622
	Working accidents	Number / CB	1.42E-05	1.17E-05
	Occupational diseases	Number / CB	5.98E-07	4.91E-07
	Toxicity potential	Points / CB	90.8	86.3
	Wages	PPP Dollar / CB	6.63	8.06
	Professional training	h / CB	1.82E-03	2.83E03
	Organization	Normalized	1.00	0.87
Consumer	Residues in Food&Feed	Rating	1.00	1.00
	Residues of GMO in food	Rating	0.03	0.03
Local/National Community	Access to land	EUR / CB	20.77	20.31
	Employment	Hours / CB	1.94	1.78
	Gender equality	%dev	44.15	42.04
	Integration	Working yrs. / CB	0.00	0.00
International community	Imports Devel. Countries	EUR	−1.79E +09	−3.50E +09
	Fair trade	EUR / CB	0.00	0.00
Future generations	Trainees	h / CB	5.50E-05	8.00E-05
	Social security	EUR / CB	4.98	8.83

Communautés Européennes' (NACE, general nomenclature of economic activities in the European Community), an initiative that classifies all industries into different sectors, or the ISIC, the International Standard Industrial Classification. All products can be linked to these NACE/ISIC codes, using the product classification list (CPA = Classification of Products by Activity).

The integrated impacts of social indexes in the Iowa farming community yielded a 57% increase in sustainable performance. Key drivers were:

(a) Days spent on professional training increased 29%; training days per year * FTE (full time equivalent) increased 56% since rationalization and mechanization on farm require a more skilled work force.
(b) Higher attractiveness of agriculturalist as a profession: In 2010 there were 20% more post-secondary students in Iowa studying agriculture than in 2000.
(c) More female farm proprietors: In 2010 about 8% of Iowa farm proprietors were female; this is a 36% increase over 2000 (5.85%).

A graphical representation of the social fingerprint of both corn production schemes is given in Fig. 11.2.

The results in Fig. 11.1 show the individual scores of the Ecology, Society and Economy of the AgBalance™ study. The Year 2010 shows better results compared to 2000 due to the normalized value being lower. In these graphs, the better score is closer to 0.6 and a worst score is closer to 1.4. These are established based on the normalized values being centered at 1 or the individual normalized value being divided by the average score of both alternatives. The Total Score graph shows the sum of the Ecology, Society and Economy assessments with each having equal weighting of 33.33%.

At the highest aggregated level, the results of the environmental, social and economic assessments are presented as single score diagrams (Fig. 11.2). The normalized values from the environmental, social, and economic fingerprint are aggregated into a single relative score through the use of relevance, societal factors and the E/C or S/C scaling factors [10]. Given that the analysis features multiple criteria and a plethora of single results, it is vitally important to show the final conclusions in an aggregated way.

11.3 Conclusions and Future Developments

Social indicators as part of AgBalance™ means integrating social parameters into the assessment model, taking all three pillars of sustainability into account, as originally proposed in the definition of sustainability by the UN Brundtland Commission. The strength of a life cycle approach is that the social aspects are evaluated along the life cycle or a defined life cycle. The assessment of social indicators shows the sustainability risks or weaknesses, as well as strengths of any given alternative. In contrast to most social sustainability assessment schemes, which serve predominantly as risk management tools, the social sustainability indicator system of

AgBalance™ aims to guide continuous improvement efforts within the agri-food value chain [4]. The UNEP/SETAC recommendations have shown to provide a useful framework to design a system that takes various stakeholder needs into account. The low attractiveness of agriculture as a profession in most if not all geographies of this world make a detailed analysis of the social situation of famers and other players in the agri-food value chain indispensable. Therefore, whilst the social sustainability indicators in AgBalance™ are currently under revision, a switch to a risk management system is not planned. However, sustainability issues caused by the increasing consolidation and globalization of agri-food value chains need to be taken into account and will need to be reflected by a revised indicator system. In particular, indicators focusing more on sustainability issues of smallholder communities deserve a more dedicated approach [11].

The AgBalance™ case study demonstrated that Iowa corn farmers improved the sustainability of their operations by an average of 40% in the decade ending in 2010. The key drivers for this advance in sustainability performance include:

- integrated on-farm innovations, namely new crop production technologies (i.e. stacked biotech traits, state-of-the-art insecticide chemistry) as well as conservation practices in management of land, such as conservation reserves (CRP) and the prevalence of conservation tillage
- farm enterprise contributions to local and state commerce, government, and education.

In sum, this case study underlines the paradigm of sustainable intensification: By adopting latest innovations, and applying conservation management and investing into the current and future workforce, intensification of an agricultural system can result in an increased sustainability.

This case study was critically reviewed by National Sanitation Foundation [9].

References

1. United Nations Environment Programme and Society for Environmental Toxicology and Chemistry. Guidelines for social life cycle assessment of products. Paris; 2009.
2. Kölsch D, Saling P, Kicherer A, Grosse-Sommer A, Schmidt I. How to measure social impacts? A socioeco-efficiency analysis by the SEEBALANCE® method. Int J Sustain Dev. 2008;11:1–23.
3. Schmidt I, Meurer M, Saling P, Reuter W, Kicherer A, Gensch CO. SEEBALANCE® managing sustainability of products and processes with the socio-eco-efficiency analysis by BASF: Greener Management International; 2005.
4. Frank M, Schöneboom J, Gipmans M, Saling P: Holistic sustainability assessment of winter oilseed rape production using the AgBalanceTM method – an example of 'sustainable intensification'?, In: Corson, M.S., van der Werf, H.M.G. (eds.), Proceedings of the 8th International Conference on Life Cycle Assessment in the Agri-Food Sector (LCA Food 2012), 1–4 October 2012, Saint Malo, France. INRA, Rennes, France, 2012; p. 58–64.
5. Landsiedel R, Saling P. Assessment of toxicological risks for life cycle assessment and eco-efficiency analysis. Int J LCA. 2002;7(5):261–8.

6. Field to Market. 2012. National report on agricultural sustainability. http://www.fieldtomarket. org/news/2012/field-to-market-releases-national-report-on-agricultural-sustainability/. Accessed August 2017.
7. Iowa State University 2001: 2000 Iowa cost and returns, FM-1789.
8. Iowa State University. 2011: 2010 Iowa cost and returns, FM-1789.
9. NSF. 2013. Submission for verification of AgBalanceTM analysis under NSF protocol P352, part B. https://www.nsf.org/newsroom_pdf/BASF_Corn_Production_AgBalance_Study_Veri fication_Feb2013.pdf. Accessed February 2018.
10. Kicherer A, Schaltegger S, Tschochohei H, Ferreira Pozo B. Combining life cycle assessment and life cycle costs via normalization. Int J Life Cycle Assess. 2007;12:537–4.
11. Frank M, Fischer K, Voeste D. BASF: measurability – A prerequisite of shared value creation in agriculture. In: Heur M, editor. Sustainable value chain management, CSR, Sustainability, Ethics & Governance. Cham: Springer; 2015.